MONITORING OF WATER QUALITY

Related publications from Elsevier

Books

Becker
Aquatic Bioenvironmental Studies
ISBN 0 444 88653 2

G. Desmet et al.
Freshwater and Estuarine Radiobiology
ISBN 0 444 82533 9

Cushing, Cummins and Minshall
River and Stream Ecosystems
ISBN 0 444 88822 5

Doerffer
Oil Spill Responses in the Marine Environment
ISBN 0 08 041000 6

Helsel and Hirsh
Statistical Methods in Water Research
ISBN 0 444 88528 5

Isaac and Shuval
Water and Peace in the Middle East
ISBN 0 444 81464 7

Jøgensen and Gromiec
Mathematical Submodels in Water Quality Systems
ISBN 0 444 88030 5

Ketchum
Estuaries and Enclosed Seas
ISBN 0 444 41921 7

Margalef
Limnology Now
ISBN 0 444 89826 3

Michael
Managed Aquatic Ecosystems
ISBN 0 444 42517 9

Taub
Lakes and Reservoirs
ISBN 0 444 42059 2

Tolgyessy
Chemistry and Biology of Water, Air and Soil
ISBN 0 444 98798 3

Vollenweider
Marine Coastal Eutrophication
ISBN 0 444 89990 1

Walker and Livingstone
Persistent Pollutants in Marine Ecosystems
ISBN 0 08 0418784 0

Journals

Agricultural Water Management
Journal of Contaminant Hydrology
Journal of Hydrology
Marine Environmental Research
Marine Pollution Bulletin
Ocean and Coastal Management
Water Policy
Water Quality International
Water Quality Newsletter
Water Research
Water Science and Technology

MONITORING OF WATER QUALITY

The contribution of advanced technologies

Edited by

F. COLIN

NANCIE International Water Centre, Vandoeuvre, France

and

P. QUEVAUVILLER

Directorate XII, Commission of the European Communities, Brussels, Belgium

Proceedings of the European Workshop on Standards, Measurements and Testing for the Monitoring of Water Quality: the Contribution of Advanced Technologies
Nancy, France, 29-31 May 1997

1998
ELSEVIER
Amsterdam – Lausanne – New York – Oxford – Shannon – Singapore – Tokyo

ELSEVIER SCIENCE Ltd
The Boulevard, Langford Lane
Kidlington, Oxford OX5 1GB, UK

First edition 1998

Library of Congress Cataloging in Publication Data

Monitoring of water quality : the contribution of advanced
technologies / edited by F. Colin and P. Quevauviller. -- 1st ed.
p. cm.
Includes bibliographical references.
ISBN 0-08-043340-5 (hc). -- ISBN 0-08-043203-4 (pbk.)
1. Water quality--Measurement. I. Colin, F. II. Quevauviller, Ph.
TD367.M66 1998
628.1'61--dc21 98-24452
CIP

ISBN: 0-08-043340-5 (Hardbound)
ISBN: 0-08-043340-5 (Paperback)

∞The paper used in this publication meets the requirements of ANSI/NISO Z39.48-1992 (Permanence of Paper).

Printed in The Netherlands.

TABLE OF CONTENTS

DISCUSSION PAPERS : USER'S VIEW

I. PREFACE TO THE REPORT AND PUBLICATION

F. COLIN and P. QUEVAUVILLER

Water can be considered as one of the basic elements supporting life and the natural environment, a primary component for industry, a consumer item from human and animals and a vector for domestic and industrial pollution. Various European Directives already provide for a framework for the control of aquatic substances, the quality of bathing, surface and drinking waters and effluent control.

Such regulatory measures are closely related to analytical measurements. However, in order to comply with these regulations on a permanent basis, treatment plants and distribution companies, as well as water users and polluters, need to implement automated measuring techniques, such as sensors and other analytical tools in continuous or sequential mode to obtain suitable alarm systems and facilitate the management of water resources and decision making processes.

This approach is in continuous evolution within the European Union. In view of the consequences of wrong measurements, it is vital that reliable quality control systems be achieved and maintained. At present, however, only a small range of analytical parameters can be measured automatically, satisfying the required quality and sensitivity criteria within an acceptable cost. It is therefore necessary to develop and validate new methods to extend the list of parameters. In addition, straightforward extrapolation of laboratory measurements is insufficient to meet the requirements of continuous monitoring. Finally, the development of new methodologies requires that the most recent knowledge in fields such as optics, electrochemistry, biochemistry, chemometrics and others be shared by experts.

In order to identify possible actions to be undertaken in the field of standards, measurements and testing for monitoring water quality, the Standards Measurements and Testing (SMT) Programme of the European Union organised a European workshop that was held in Nancy on 29-31 May 1997. This Workshop brought together researchers and industrial users with the aim of making an overview of the present state of the art, of considering possible improvements in existing techniques and the need and possibilities of developing new advanced technologies.

A Steering Committee prepared the programme and, in close consultation with the European Commission's SMT Programme, identified experts in the above mentioned fields who were invited to submit written contributions to the programme. This intensive 2-day event in Nancy, was organised and steered by the International Water Centre, NANC.I.E.

The following themes constituted the framework of the seminar :

- Diagnosis :
  - End users views
  - Identification and prioritisation of research needs

- Advanced technologies for classical parameters: improvement of existing methods or development of alternative techniques
- Advanced technologies for new parameters and measurement concepts
- Needs and possibilities for the development of advanced technologies: establishing priorities.

The Workshop programme was organised in such a way as to examine technologies adapted to different water categories :

- natural surface and underground waters;
- drinking water;
- water for industrial use and process water;
- domestic and industrial waste waters; and
- sludge arising from water treatment processes.

Also, it was structured to match the needs and the opportunities arising from quickly evolving basic sciences and measurement technologies :

- optical technologies
- electrochemical sensors
- biochemical sensors (whole cells: bacteria, micro-algae...)
- microelectronics and computer sciences for on-line signal acquisition and treatment
- chemometrics for the interpretation of data treatment (numerical, neuronal, fuzzy techniques)

It became evident in the preparation of the meeting and professional contacts established, that due to the spatial and time dependent variability of water characteristics, on-line monitoring is needed. Yet, it became evident that corresponding techniques are rather limited in terms of the parameters covered and sensitivities available compared to existing laboratory techniques. At the outset of the Workshop, the participants agreed that a continued research effort was necessary to improve on-line monitoring performance and to encourage wider acceptance and promote their use.

Words of appreciation have to be expressed to the experts whose papers are reproduced in this publication and Jacek Zapasnik, European and Multilateral Affairs Manager of the NANC.I.E. staff who organised the seminar and acted as facilitator during the plenary and round-table sessions.

The proceedings of this Workshop are not only published to reflect the issues raised during this event, or to merely identify priorities for the development of advanced technologies in water quality monitoring. The Workshop gave a privileged opportunity to exchange experiences and develop certain guidelines, but more importantly to establish a framework for cooperation on the European level between professionals involved actively in this field.

II. PROJECT SUMMARY

1. Background

Various European Directives provide already a framework for the control of aquatic substances, the quality of bathing , surface and drinking waters and effluent control. Such regulatory measures are closely related to analytical measurements. However, in order to comply with these regulations on a permanent basis, treatment plants and distribution companies, as well as water users and polluters, need to implement automated measuring techniques, such as sensors and other analytical tools in continuous or sequential mode to obtain suitable alarm systems and facilitate the management of water resources and decision making processes.

2. Objectives

It is vital that reliable quality control systems be achieved and maintained. At present, however, only a small range of analytical parameters can be measured automatically, satisfying the required quality and sensitivity criteria within an acceptable cost. It is therefore necessary to develop and validate new methods to extend the list of parameters. In addition, straightforward extrapolation of laboratory measurements is insufficient to meet the requirements of continuous monitoring. Finally, the development of new methodologies requires that the most recent know-how in fields such as optics, electrochemistry, biochemistry, chemometrics and others be shared by experts.

3. Work programme

In order to identify possible actions to be undertaken in the field of standards, measurements and testing for monitoring water quality, the Standards Measurements and Testing (SMT) Programme of the European Union, organised a European workshop that was held in Nancy 29-31 May 1997. This Workshop brought together researchers and industrial users with the aim to make an overview of the present state of the art, to consider possible improvements in existing techniques and the need and possibilities of developing new advanced technologies.

4. State of progress

It became evident in the preparation of the meeting and professional contacts established, that due to the spatial and time dependent variability of water characteristics, on-line monitoring is needed. Yet, it became evident that corresponding techniques are rather limited in terms of the parameters covered and sensitivities available compared to existing laboratory techniques. At the outset of the Workshop, the participants agreed that a continued research effort was necessary to improve on-line monitoring performance and to encourage wider acceptance and promote their use. The proceedings of this Workshop will not only be published to reflect the issues raised during this event, or to merely identify priorities for the development of advanced technologies in water quality monitoring. The Workshop gave a privileged opportunity to exchange experiences and develop certain guidelines, but more importantly to establish a framework for cooperation on the European level between professionals involved actively in this field.

III. CONCLUSIONS

1. Present situation and needs

1. 1. General remarks regarding all categories of water

The functioning of several existing types of monitoring equipment poses a problem. It results not only in a lack of reliability of the equipment itself but also in problems of fouling. This state of matters is likely to affect the development of the market.
There is need for a validation of sensors and analysers functioning under field conditions. For this, specific methodologies (procedures, test benches) are to be developed bringing together realistic conditions of evaluation and reproducibility of these conditions over a period of time.
The development of new equipment suffers from a problem relating to the cost of final stages which correspond to proportions of 1/10/100 respectively for research, development and industrialisation. These costs are all the more difficult to support when the market is limited and the development is carried out on national scales.
Toxicity monitoring poses specific problems particulary with respect to the interpretation of results.
The measurement step itself in not the only aspect to be considered, as sampling and the presentation of results are just as important. In each case it is necessary to start by answering a few questions. Why is there a need to measure? Where are the measurements to be obtained? What are the quantities to be measured? What time resolution is required? It is not necessary to measure all the relevant quantities in real time, but a selection must be made taking into consideration the time constant of the process or the phenomenon studied.
A wide gap exists between field research results of sensor technology and effective implementation by the end user. Several innovative prototypes have been developed on a laboratory scale to deal with traditionally measured quantities, however, only a small number of them reach the market.

This can result from :

- Inadequate communication between research workers and users. The use of a common vocabulary could be a starting point.
- A lack of confidence of the users in the effective functioning of samplers ans analysers.
- A gap between the required sensitivity and the level effectively achieved.

One should not strive to develop identical equipment for "small" and "big" users (size of the installation, technical levels).

It is necessary to rank the monitoring according to their status: those which are designed and developed, those which are validated and finally those which are commonly used in the field.

1. 2. Specific remarks applying to different categories of water

One may consider in general and to simplify matters two major types of water, taking into account similarities in needs and/or type of apparatus used:
Water that is (relatively) clean natural water, potable water, water for industrial use, certain types of water in industrial processes (e.g. cooling water).
Water that is highly polluted: urban or industrial wastewater, sewage sludge.
The identification and ranking of needs can be take place at the level of each of these two main categories:

Table 1 : Characterisation of clean water : The present situation concerning a certain number of species of major interest

Species	Laboratory analysis*	Field (**)	Remarks
Cl_2	+++	+++	Miniature sensors are needed for field measurements
O_3	Not applicable	+++	Air monitors currently markeded may be used although they are expensive
BrO_3^-	++	0	Difficult to measure in the lab though currently feasible
$C1O_3^-$	++	0	
NO_3^-	+++	++	uv-abs. measurements require very clean waters
HPO_4^2	+++	++	Titrator colorimetric units ; require absence of particles
NH_4^+	+++	+	Low level analysors (in-situ) do not work properly
Cr^{III}/CR^{VI}	++	0	Speciation will be essential in the near future, still needs laboratory development and instrument Maybe it will not be needed " in-situ "
VOC_S	+++	+	Electrical sensor arrays look promising ; it's possible to use SERS to monitoring individual chemical species, and not VOCS as a whole. Immunoassays are certainly possible, but still needed the proper antibodies. Promising by IMS.
Drug metabolities	+++	0	Immunosensors are possible, with targeted antibodies
Pesticides	+ (+)	+	" Strip " immunoassays for field analyses are available, with precision similar to that reached in the lab for certain pesticides ; immunosensors are promising
Desinfection by products (DBP_S)	+	(0)	They need still to be defined (organochlorides, chlorophenols, BrO3- etc.)
CO_2 Hardness	+++	++	May be necessary in certain areas where water is preconditionned ; already being measured in those areas

(*) Keys :
+++ : well developped
++ : need some " tuning "
\+ : still in its infancy
0 : not available yet

(**) refers to both sensors and miniaturized total analysers systems

1. 3. Characterisation of waste water

Table 2 : The present situation concerning a certain number of species of major interest in urban waste water

Parameters	laboratory analysis*	Field (**)	Remarks
pH	+++	++	Need of ruggedness, autocleaning, auto-calibration
rH	+++	++	Need of automatic cleaning, prevention of electrode contamination
O2		+++	Need automatic cleaning and automatic calibration
H2S	++	+	In sewers, very harmful
TSS at inlet	0	+	Indirect measurement via turbidity, need automatic cleaning of detritus, grease
TSS at discharge	0	+	Need of cheap sensors with auto-cleaning, without maintenance
Settlability of sludge	+	+	Equipment derived from lab, too complex & costly
Siccity of sludge	+	0	Field test, not on-line instrument, IR or EM radiation absorb. Possible
Sludge blanket level	no applic.	++	Optical absorb. or U.S. reflection, need auto-cleaning and sensitively
NH4+	+++	++	Not yet in-situ (off line), too costly, too much maintenance
NO43-	+++	++	Not yet in-situ (off line, too costly, too much maintenance
Total nitrogen	++	+	Not really available, too complex, too costly
PO43-	+++	+	Only by colorimetry for analysers, too costly, much maintenance
Total phosphorus	0	+	Nedd to convert P to PO4 first, not really available on-line, not effective
COD	++	+	Derived from lab, national standard, not CEN not really available on-line, too costly
BOD	++	0	Analysers far from standard (5 days)
TOC	+++	+++	Too costly, too much maintenance
Biodegrada-bility	++	0	Concept to be clarified and standardised
Toxicity	++	++	Many test, very costly, high maintenance, difficult to interpret

Keys :

+++	:	well developed
++	:	need some " tuning "
+	:	still in infancy
0	:	not yet available
*	:	refers to automatic analyser system
**	:	refers to both sensors and miniaturised total analysers systems

Conclusions

The round table experts developed a double classification of priorities: at the level of fields of application and at the level of parameters to be monitored. They arrived at the following situation where " * " indicates priority for real-time monitoring.

Areas :

- environmental quality :
  - evolution/modelling
  - monitoring physical, chemical and biological processes (*)
- treatment process :
  - feed water alarm (*)
  - process control (*)
- distribution system and transport (control of quality degradation) :
  - drinking water (*)
  - industrial water (*)

The following parameters were considered as being important with respect to the above mentionned areas, where the " + " indicates the priority for improving methods or developing new techniques :

Parameters :

- Physical :
  - turbidity (+)
  - conductivity
  - temperature
  - pH
  - redox
  - particle number concentration (+)
- Global :
  - toxicity (+) (note : toxicity with respect to drinking water refer to : continous monitoring system of global water quality with respect to human health.
  - TOC (+)
  - BDOC (+)
  - humic acids
  - VOC (as a global parameter)
  - taste and odour
- Organic :
  - DBP's (+)
  - pesticides (+)
  - PAH's
  - pharmaceutical residues
  - vinylchloride
  - acrylamide
  - VOC (es specific compounds)
  - surfactants

- Inorganic :
  - sulphate
  - chloride
  - phosphate (+)
  - nitrate (+)
  - ammonium (+)
  - hardness
  - alkalinity
  - aluminium
  - heavy metals
  - speciation (i.e. Cr, As, Se) (+)
  - chlorine (free redidual) (+)
  - nitrite
  - lithium
  - ozone (+)
  - bromate (+)
  - lead
- Microbiological :
  - coliforms (fast method) (+)
  - giardia and cryptosporidium (sampling)

The COD is not standardised at the European level and there is a lack of relation between COD and alternative parameters such as total organic carbon (TOC).

COD, total suspended matter (as such, and not as evaluated by traditional measurement of turbidity), total nitrogen and total phosphorous are not accessible to continuous monitoring.

New concepts of measurement are necessary to determine COD, BOD and TOC - the techniques and methodologies developed must be correctly defined.

One does not have suitable concepts and parameters for the measurement of biodegradability in anaerobic media.

It is desirable to carry out intensive monitoring in industrial process circuits to control and prevent discharges.

2. Strategy for the application of advanced measurement technologies : improvement of existing techniques or development of alternative techniques?

After presenting a few of the advanced measurement techniques and highlighting their potential (refer to chapters III and IV), it is necessary to identify potential applications for meeting the needs expressed previously. This problem was evaluated differently for global parameters (description of behaviour) and for well defined and identified chemical species.

2. 1. Measurement of global parameters

General issues highlighted

- Complexity of global parameters / no " traceability audit trail "
- Appropriate selection of global parameters to fulfil user needs / due to complexity, rational choice of parameters difficult
- Fouling of devices (different for optical and electrochemical techniques)
- Need to relate multiple global parameter data to yield increased information - validated approach to global parameters selection using chemometrics
- Remote diagnostic / reporting state of " sensor health "

Specific issues of global parameters

- pH, Redox, conductivity : fouling
- Suspended particulate matter : Unclear relationship between turbidity and suspended particulate matter, standard turbidity methods do not fully exploit the technique
- Total Organic Load : requires clear definition BOD, COD etc only give an indication of TOC lack of suitable on-line methods
- Toxicity : must relate measurement result to specific needs, e.g. for waste or drinking water, alarm systems on-line measurements, low cost, sub-classes toxicity
- Odours : need to relate measurement to process control regulation / difficult to standardise
- Global parameters to be measured otherwise : BOD, COD, TOC
- Other Global parameters desired :
  - Total heavy metals
  - Nutrients
  - Halogenated compounds
  - Total hydrocarbons
  - Pathogens and microbiological contaminants on-line (time dependency of global parameters useful, e.g. BOD)

2. 2. Measurement of individual chemical species

6.2.1. A first point that was discussed concerned the choice between multiple component analysis and highly specific analysis. The strengthening of regulations will induce the need for a greater specificity. These two approaches are not mutually exclusive as shown in the three examples below :

Spectrometric analytical methods such as atomic absorption, plasma optical emission spectrometry or coupling of plasma torch and mass spectrometers can be installed on-line (subject to miniaturisation and adaptation of the technology).

Use of immunological techniques allows the rapid detection of specific families of pollutants (e.g.: PAH, PCB...) but not of individual species.

Chemometrics can be used for increasing specific information extracted from multiple measurements having a more global character (multiple component analysis).

In all the cases, fouling may be a crippling problem. There is then an urgent need for co-ordinated research regarding the prevention or solution of problems related to fouling for all different types of online monitoring and measurements.

6.2.2. A pre-separation integrated in the measurement system can improve the selectivity of the measurement and by this means its significance for the nature of sample. Pre-separation, however, can disturb the precision of the measurements. The possibilities for pre-separation are dependent on the nature of the water to be analysed and on the time of frequency at which the measurements have to be carried out : on line, discontinuous, instantaneous etc. In all cases, the most critical operations in the complete analytical method - i.e. sampling, pre-concentration, pre-separation - must be subjected to validation, which requires the development of corresponding procedures.

It is difficult to carry out measurements that strictly continuous and instantaneous but this is generally not necessary (except in the case of drinking water, concerning bromate and organochlorides)

The choice depends on the intended application and particularly on the dynamics of the process monitored.

6.2.3. With respect to validation, no notable work seems to exist. The validation procedures and criteria must be established for each case, depending on the working principle of the monitoring apparatus. Validation needs are expressed both at field and laboratory levels.

6.2.4. It is necessary to carry out pilot projects on a demonstration basis to convince end users of the advantages new techniques (sensors, miniaturized analysers) have for on-line monitoring. Resources are needed for conducting such long-term projects. Actions are underway for measuring dissolved oxygen. There are needs regarding other chemical species.

6.2.5. Highly specific techniques have been introduced, thet may be solution for the demand in-situ and/or on-line of field analyses of individual chemical species:

- miniturized ion mobility spectrometers (IMS), coupled to compact separation systems
- surface enhanced Raman scattering (SERS)
- laser diode atomic absorption spectrometer (LD-AAS)
- optical sensors (optodes).

Other techniques were also mentionned in the Workshop :

- electrochemical sensors and biosensors
- microbiological sensors
- biosensors.

2. 3. Application of data processing techniques

Because of their generic character, multivariate techniques may be applied to all measurement results with a special interest for the characterisation of water behavior (properties having a global character).

6.3.1. The situation is characterised by a large variety in the form of measurement data to be processed:

- specific sensors and sensor array (less specific), redundant sensors
- continous or discrete data ?
- constant sampling period ?
- heterogeneous data
- noise
- time series
- data without time reference
- data with superposed effects
- information given by the operators (qualitative ou semi-quantitative)
- minimal sensors and soft sensors

6.3.2. The techniques to be used are determined by the measurement objective and information to be obtained. Here are some of these:

- to monitor the process
- to detect changes in the process
- to follow the process evolution
- better understand the process
- to analyse sensor functioning

6.3.3. Depending on each case, responding to these objectives will face difficulties:

- global treatment is impossible
- local treatment is possible
- data continuity during the time
- data storage

6.3.4. For solving these problems, research efforts and in particular the transfer of knowledge and skills (applied mathematicmetrics) are necessary in the following fields:

- tools for data validation
- signal and image processing
- standardization of protocols
- system modeling
- design of sensor networks
- sensor modeling
- estimation of missing data
- maintenance and calibration of sensors
- release correct information to the operators

3. Towards an overall synthesis and immediate recommendations

The seminar allowed the updating of a list of needs in the field of monitoring the water quality that presently cannot be satisfied or only poorly satisfied and provided the state of the art (summary) regarding currently available means. These needs have been identified and ranked. Several advanced measurement technologies of physico-chemical, chemical, biochemical, biological nature have been developed at the laboratory level and seem to have interesting potential :

- Several technologies were well covered and illustrated during the workshop:
  - Optics
  - Microbiology
  - Electrochemistry for VOC measurement in gas phase

- Because of lack of time, other technologies which were just as interesting could not be studied:
  - Electrochemical techniques in general (aqueous phase)
  - Separation techniques
  - Micro-electronics
  - Materials
  - Miniaturized analytical systems

4. Recommendations

4. 1. First recommendation

As a result of the above, it was agreed to pursue the objectifs identified during this workshop.

For this, it is proposed to carry out a study, a number of research topics, which should :

- provide additional details of needs (parameters to be measured, limits of detection necessary, degree of acceptable uncertainly etc.)
- have the objective of being more exhaustive, detailed and justified by quantitative and reference elements

This study must extend the scope and examination of possible contributions of advanced measurement technologies (notably those which were not examined in detail during the present workshop), basing itself on:

- all available scientific and technical literature
- research work conducted under specific European Union programmes (SMT, ENVIRONMENT AND CLIMATE, ESPRIT, BRITE EURAM, EC MAST) with the perspective of valorization of their results.

Finally, the study must include the following points :

- Analysis of limitations to transfer to real application, using selected examples on the basis of information given above.
- Recommendations for removal of limitations:
  - technological aspects
  - regulatory aspects
  - institutional aspects

4. 2. Second recommendation

The problem of validating on-line and in-situ monitoring systems for the quality of water appears to be a key element in the dissemination of new techniques. Faced with the lack of suitable validation methodologies and protocols, it is proposed to prepare a collective research project oriented towards the most delicate type of monitoring technique.
The title of this project could be: "Validation in real conditions of new water quality monitoring systems with biological respose and related physico-chemical measurements".

4. 3. Third recommendation

It is proposed the the definition of future research programmes should consider on a priority basis all the problems limiting the effective application of new measurement and monitoring technologies arising from research.
Priority must be given to more general problems (notably the fouling of measurement devices) and to research on generic solutions (instrumentation, data processing...) likely to be applied in all the fields related to the monitoring of water quality.

4. 4. Fourth recommendation

Participants of the Workshop expressed their wish to pursue their contacts in a formal or informal manner, favouring a pluridisciplinary approach and dialogue between researchers and users in Europe.

VI. PAPERS

OPTICAL TECHNIQUES FOR WATER MONITORING

P. SCULLY, School of Electrical Engineering, Electronics & Physics, Liverpool John Moores University, Byrom Street, Liverpool, L3 3AF. United Kingdom.

1. Abstract

A review of optical techniques used for water monitoring is presented, including sensors based on optical fibres, waveguide and integrated optics technology. The impact of the optoelectronics revolution in reducing cost of optical components and stimulating new materials and devices, is discussed, along with new concepts in sensor technology such as biosensors and immunosensors based on optical techniques. The effect of legislation versus cost of new technology is considered. Finally, examples of optical sensors suitable for water monitoring are described, including commercially available sensors.

2. Introduction

Classic optical techniques for chemical analysis have been traditionally used for water quality monitoring and trace chemical analysis; e.g. UV-Visible spectrophotometry, IR-spectrometry and spectrofluorimetry. The characteristic transmission, absorption or fluorescence spectrum of a chemical species is measured in order to determine its concentration or identity. In addition, techniques such as Microtox for toxicity measurement, exploit bioluminescent bacteria affected by toxins, to generate light. Such instruments are big and bulky, cost in the region of Stg 20 K, and require the sample to be taken to the instrument, which is operated by a highly skilled technician.

The burgeoning, multi-billion dollar optical fibre communications market, expanding throughout the Eighties, with associated mass production of optical and optoelectronic devices, has lowered the cost of existing optoelectronic devices, and enhanced R&D interest in new optoelectronics, light sources, detectors and new optical materials. This has paved the way for a new generation of optical sensors and measurement techniques, which exploit the new, low cost and compact light sources, detectors and devices, arising from the optical communications revolution. An excellent example of the fruits of the Optical Era, is the Compact Disc player, which has spiralled in price over a decade from 100s of Stg, to 10s of Stg. This consumer item, owned by every European teenager is as much a symbol of the Optical Age, as the Transistor Radio was of the Electronics Age of the 1960s.

Existing optical and spectroscopic techniques can be adapted, using new optical components, to form new, miniaturised, low-cost and portable sensors and optical instruments for applications in medicine, process control and for the water industry. New optical techniques, materials and devices are continually evolving due to the optical telecomms market driven research, enabling generation of new optical measurement and sensing systems.

3. Optical sensor and measurement systems

Due to the diversity of techniques and technologies available both commercially and in research labs and reported in the literature, it is difficult to classify optical sensing and measurements. Different classification schemes have been used, ranging from those based on the physical or chemical quantities measured by the system, to those based either on the physics of the sensing mechanism, or the detection system, or the modulation of the light [1,2,3,4,5].

An optoelectronic system can be divided into its component parts :

- The light source
- The light detector
- Coupling optics

The coupling optics can either be based on conventional lenses, optical fibres or waveguides. By using integrated optics, all three components can be placed on the same chip or substrate.

3. 1. Optical fibre sensors

Optical fibre sensors form a large sub-set of the family of optical sensing and measurement techniques, and are particularly relevant to the water industry because of the ability to perform point and distributed measurements remotely from the interrogating electronics.

Work on optical fibres was initiated in the 1960s, and optical fibre sensors was one of the spin-off technologies. The technology is sufficiently mature to be the subject of several review books [2,3,4,5].

In an optical fibre sensor, the fibre forms the coupling optics, and transmits the light from the light source, to the modulation zone, where the properties of the light are modulated in response to a change in an external parameter, which can be physical, chemical or biological. The light is then transferred to the detector, where the perturbation in the light characteristics is converted into an electrical signal.

The advantages of optical fibre sensor systems over conventional sensor systems have been well documented and are summarised as follows :

- Immunity to electromagnetic interference.
- Electrical isolation compatible with intrinsic safety requirement, rendering such sensors to be useful in wet environments.

- Passive operation, ensuring no power or electrical circuits are required at the sensing point.
- Transmission of light over long distances, enabling remote or distributed sensing due to the low losses achievable in optical fibres.
- Integration of sensing and telemetry in intrinsic optical fibre sensors.
- Chemical immunity to corrosion enabling use in hostile environments.

Optical fibre sensors can be categorised as [6] :

- Intrinsic sensors in which there is no break in the fibre, and the characteristics of light carried within the fibre are modified by the measurand acting upon the physical properties of the fibre.
- Extrinsic sensors in which the light leaves the fibre and enters a physically separate modulator, where the light characteristic are changed by the measurand.

The advantages of extrinsic sensors are that the modulator can be specially designed to maximise the modulation of the light signal. The disadvantage is that the number of optical interfaces is increased, with a resultant loss of optical power and the danger of contamination of optical surfaces.

Optical fibre sensors can also be classified according to the characteristic of light that is modulated by the measurand: intensity, wavelength, polarisation and phase.

3. 2. Optical fibre for optical sensing

Optical fibre sensors came into being because of the development of optical fibres in the 60s. Several types of optical fibres are currently available, based on silica glass, which can be manufactured in the form of single mode and multimode step index and graded index fibre. These fibres are now cheap due to large scale manufacture, resulting from the optical telecomms industry.

3.2.1. Optical fibre may be classified as

Single Mode fibre, which allows the propagation of one or few modes in the fibre, and which require coherent laser light sources to excite the fibre. Polarisation maintaining fibre are a subset of monomode fibres, and such fibres form the basis for intrinsic optical fibre sensors using phase, polarisation and evanescent field modulation. The problem with such fibres is that they do not carry much light.

Multimode fibres allow propagation of 100s or 1000s of modes and therefore are much better at coupling light from broad band incoherent light sources such as tungsten halogen lamps, LEDs and multimode laser diodes, to the modulation zone, and then back to the detector. Such fibres form the basis for extrinsic sensors, and in particular for fibre optic chemical sensors (FOCS), which are primarily of interest to the water industry, because they exploit chemical reactions sensed in the visible and UV part of the spectrum.

3.2.2. Optical fibres made of materials other than silica are now available and can be used for sensing

Plastic Optical Fibres (POF) are thought to be the poor cousin of silica based glass fibre in terms of their poor transmission properties (20 dB/km instead of 2 dB/km) making them only suitable for Local Area Networks (LANS) and crude sensor applications. However, recent advances in polymer materials in Japan [7,9], leading to low loss and high bandwidth, graded index POF, with improved attenuation characteristics and increased temperature tolerance, is posing a serious challenge to glass fibre. Such fibres are far easier and cheaper to terminate than silica glass based fibres, being easy to cut and polish. They usually have 1 mm diameter core, which are easy to couple light into, and requiring low tolerance, injection moulded plastic couplers. Fluorescence doped plastic optical fibres are a new venture in POF, in which the core is doped with fluorescent dyes. When such fibres are excited by a suitable wavelength, fluorescence occurs at a longer wavelength, and the fluorescent light is guided within the fibre core. Such fibres act as light detectors, amplifiers and wavelength shifters and their applications in sensing are becoming apparent [7].

Polymer clad silica fibres (PCS) are large diameter, step-index, multimode silca fibres, with a polymeric coating that can be chemically removed, exposing the cladding. Such fibres form the basis for multimode evanescent sensors, widely used as Opical Fibre Chemical Sensors [8].

UV and IR transmitting optical fibres :Conventional silica fibres are damaged by, and are highly attenuating at UV wavelengths, and their use is restricted to the 300 to 1800 nm wavelength region, which limits their use in sensors based on UV or IR spectroscopy. Fibres transmitting below 230 nm have been developed, with a transmission of 0.3 dB/m at 210 m, enabling UV light sources such as deuterium lamp, to be used for fibre based spectroscopic applications [11,12]. Fibres have also been developed for the mid-IR region, based on fluoride (0.5 to 5 μm), chalcogenide (1 to 6 μm) and silver halide (3 to 15 μm).

3. 3. Light sources for optical sensing

A major development, also driven by the optical telecomm unicattons industry, is the semiconductor laser diode or injection laser, producing monochromatic, coherent light at infra-red and red wavelengths. These were initially developed to match the transmission windows in optical fibres at 1.3 and 1.55 μm, but are now commercially available at near infrared (950, 850, 780 nm) and red wavelengths (660, 630 nm). Together with new developments in LEDs, such light sources have been mass produced at low price and are optimised for optical fibre coupling (pigtailed lasers etc.). They have the advantages of low power consumption, high stability and long lifetime, as well as being compact, robust and easily intensity modulated.

The availability of cheap laser diodes have fuelled interest in associated laser diode driven light sources, such as fibre lasers, and diode driven solid state lasers, giving an even greater range of wavelengths and intensities, and the possibility of using new optical materials and crystals for frequency doubling, and frequency mixing . Advances in semiconductor technology have led to green and blue LEDs and laser diodes being commercially available, as well as the possibility of compact, cheap UV lasers [13,14].

Such light sources would be of great importance to the water industry, which relies on laboratory based UV-visible and fluorescence spectrometers and colourimetric tests, with inherent bulky, heat generating incandescent filament bulbs, interference filters and monochromators, or bulky, expensive, cooled, three phase driven visible and UV lasers (Argon-Ion, eximer), because they would enable compact, on-line fibre based spectroscopy systems to be developed. LEDs are both longer lived and more efficient than incandescent bulbs, with a lifetime of up to 15 years, but a power consumption of just a tenth.
Tuneable lasers over the range 200 to 10,000 nm, can be constructed using optical parametric converters and difference frequency generation, and have important applications in spectroscopy. Diode pumped solid state lasers (DSP) and frequency doubled DSPs are replacing helium neon and argon ion lasers [15,16].

Electroluminescent films (ELF), usually used for backlighting of displays, are available in a range of wavelengths, with high stability, and can be used to excite fluorescence in optical sensors. ELFs emit a diffuse pattern of light over their surface, and can be wrapped round glass vessels, to excite fluorescence of a measurand liquid. [17]

3. 4. Light detectors for optical sensing

Once again, the range of light sources is driven by the optical telecomms applications. Light detectors include silicon or germanium based pin photodiodes, APDs and photomultipliers. For sensor applications based on absorption or fluorescence spectroscopy, a dispersing element is required. Compact scanning monochromators, can be used, but an exciting recent development has been fibre compatible CCD linear array spectrometers, which are now commercially available from companies such as Zeiss, Ocean Optics and Microparts, with resolution of a few nm. Such devices provide spectral information in real time, have no moving parts and are compact in size.

Formation of a grating on the end of an optical fibre, negates the requirement for a separate defractive element, thus avoiding alignment and contamination problems, and such gratings can be formed on the distal ends of plastic optical fibres using a phase mask and 248nm eximer laser light, and interrogated using a CCD linear array. Applications of fibre end gratings are numerous, including wavelength sensing and referencing and wavelength division multiplexing (WDM) [18].

3. 5. Integrated optical systems and new materials

Optical light sources, devices and detectors all on the same chip or substrate enable further miniaturisation of optical systems, including devices such as waveguides, gratings, couplers. Several sensor components have been integrated on a chip, by Texas Instruments, combining light sources, waveguides, detectors and coupling optics in a single device, compatible with a printed circuit board [15,62].

The optoelectronics industry has stimulated interest in new optical materials, such as optical polymers and photorefractive materials. Sol-gel is a material which is of particular interest for sensor applications.
A porous glass matrix produced at low temperatures (<100°C), enables chemical molecules, usually analyte sensitive dyes, to be trapped in the pores. Sol-gels were initially used to make a chemically porous glass tip for silica based fibres, but are now being used as chemically sensitive coatings for a variety of optical systems, such as thin film coatings on substrates and waveguides, and as alternative cladding materials for evanescent wave planar sensors and surface plasmon mode systems [11,20,21,22,23].

4. Water quality monitoring

The water industry has traditionally used laboratory based analytical techniques such as AAS (atomic absorption spectroscopy), GC (gas chromatography), HPLC (high performance liquid chromatography), MS (mass spectroscopy) and GCMS (gas chromatograph mass spectrometer). Optical techniques include UV-visible and IR spectrophotometry and fluorescence spectrophotometry. Such instruments are used for diagnostic measurements, in order to determine contaminants, sources of contamination, and toxic effects.

Monitoring measurements have a different philosophy, and require frequent, preselected, repetitive sampling. The use of expensive, time consuming diagnostic methodologies is not appropriate here; analytical versatility must be sacrificed for dedicated, faster and more economical systems [24].

4. 1. Legislation versus cost and realistic technology

Legalisation determines what parameters are monitored; in Europe, recent EU laws address the chemical and micro biological quality of drinking water, composition of industrial discharges, bathing water quality, waste disposal and many other issues[25].

Once a substance is implicated by legislation, sensors will be required to monitor it. However, species in water can often be determined accurately and cost-effectively by existing laboratory analytical methods. Sensors are only required if it is necessary to monitor the species at the point of discharge, or in the external environment [25]. In addition, legislators only stipulate continuous monitoring if realistic technologies exist. A very large number of chemical and other quantities are implicated by legislation, but no generic measurement techniques, sensing principles or technologies are available for their determination. Thus there are a number of barriers to be overcome, before optical techniques, developed in the research laboratories, are accepted as necessary by the water industry, and become commercially viable.

Table 1 : Parameters for in-land water, implicated by environmental legislation: After Bogue [25].

Ammonia	Nitrates
Phosphates	Metals
Halogenated organics	Suspended solids
Cyanide	Pesticides
Fuels and oils	Biochemical oxygen demand (BOD)
Chemical oxygen demand	Total organic carbon (COD)
Algae and algal blooms	Bacteria and viruses
Phenols	Chlorine
Dissolved oxygen	pH

An example of slow uptake of new monitoring technologies is illustrated by the National Rivers Authority (NRA) in the UK, which samples river water in order to measure 200 to 300 different analytes. Over 4.3 million determinations are made annually, and 95% are achieved by traditional sampling and laboratory analysis. Techniques such as AAS, GC, HPLC, MS and GCMS are widely used, and even taking into account new sensor and instrument developments, the NRA estimates that field measurements with sensors will not exceed 10% of the total [25]. This is because laboratory analysis can be very cost effective, when large numbers of samples are involved.

Trace level determinations are required (sub ppb), which exceed the detection limits of most chemical sensors.

For compliance monitoring, approved high accuracy analytical methods are stipulated by legislation.

In addition, sensors are not available for most of the analytes involved, and commercially available sensor based field measurements in the aquatic environment are presently restricted to parameters such as pH, DO, ammonia, conductivity, some metals, turbidity etc.

4. 2. The impact of new optical technology on water quality measurement

The replacement of existing sensors and monitoring instruments and techniques with new optical techniques, depends on the following factors:

The technical benefits of optical sensors over existing sensors: i.e. will they work better and be more reliable and accurate with improved sensitivity, specificity, response time and capability of measuring new parameters?
Will they offer economical benefits over existing techniques: i.e. are they cheaper.
Reducing ownership costs (5-year life costs: consumable, reagents and purchase price can total Stg 10 to 20 K). Multiparameter sensors, disposable sensors, on-line sensors also reduce cost.
Situations where sampling and laboratory analysis is too slow and/or costly.
Where legislation stipulates new monitoring regimes: i.e. species, applications etc.

One of the most obvious places to start, is to replace, simplify and reduce the size of existing optical instruments using current spectrometric and chemical techniques, to make them cheaper, more compact, and remove them from the laboratory, to the test site. The largest and oldest class of optical sensors, Optical Fibre Chemical Sensors (OFCS), were developed in this way, and are now available commercially.

Limitations of existing on-line water quality analysers for ammonia, TOC, metals and nitrates include high cost of ownership and limited field reliability.
These limitations can be overcome by new optical techniques, which will enable species to be determined that are at present, difficult or impossible to measure on-line. Such techniques include ultra-violet absorption, fluorescence spectroscopy, as well as immunoassay, enzyme reactions and live cell biosensing effects, which can be configured to produce or modulate light. Such sensors are made possible by the new light sources, fibres detectors and integrated waveguide technologies described in Section 2 [26].

To date, no on-line analysers yet exist for pesticides, biochemical oxygen demand (BOD), halogenated and other organics (TCE, phenols etc) and toxicity. Problems include the use of sensing techniques that require wet reagents, the requirement for frequent calibration, the need for often complex sampling systems, and the lack of availability of suitable sensing methods. These can be overcome by remote fibre coupled optical sensors, reagent free techniques and disposable sensors. Many of the most promising sensor principles and technologies are already being exploited commercially in medical and process control fields [26].

5. Fibre optic chemical sensors (FOCs)

Due to their exploitation of traditional spectrometric and standard colourimetric chemical techniques, FOCs form the oldest, and most well developed class of optical sensors, and can be classified into two categories [11].
Direct spectroscopic, where the spectral power distribution of the light is modulated by the spectral properties of the analyte. Using the Beer-Lambert law, concentration of a chemical species can be measured by traditional analytical techniques such as fluorescence or absorption.
Reagent mediated, where an intermediate chemical reagent reacts with the target analyte to cause light modulation such as an absorption change, fluorescence etc.

The reagent may be in a liquid form or immobilised in a solid matrix such as sol-gel. FOCs are fibre sensors with preselected chemical and/or physical properties attached to the distal end and/or sides, so that sensitive specific in-situ remote real-time analyses can be made. FOCs can be based on fluorescence , absorption, Raman effects, refraction or reflection. Interaction of the analytes with the sensing reagents produces a change in one of the mentioned spectroscopic parameters [19,27].

FOCs enable inexpensive spectroscopic sensors for specific chemicals, simplifying the spectrometer back end, and reducing the price. They consist of a light source, the sensitised fibre, a spectral sorter/disperser/filter, and signal processing and display electronics.

The wavelength or spectral modulation technique is usually derived from traditional wet chemistry analysis and absorption or colourimetric and fluorometric techniques. The appropriate chemical reaction is chosen to have the specificity, sensitivity and stability required, and a FOCs design is matched to the optical modulation properties of the reaction.

FOCs designs are as follows [24] :

- *Tip coated FOCs*
  A reagent is immobilised on the tip of the fibre, usually large diameter multimode fibre. The reagent may be contained within a membrane or polymer, attaching large diameter, porous fibre coatings (sol-gels) [11], or porous glass beads, or using highly sensitive reagents.

- *Evanescent Wave FOCs*
  Removing the lower refractive index cladding from an optical fibre, exposes the core to the analyte. Light propagating within the core of the fibre, creates an evanescent wave in the cladding, which interacts with the analytes at the core-cladding interface, causing spectral attenuation of light within the core, or fluorescence excitation generating guided waves at fluorescent wavelength within the core. Fibres can be tapered to increase the evanescent field interaction [29].

- *Side coated FOCs*
  Multilayered FOCs in which the reacting analyte is sandwiched between the core and the cladding, forming an intermediate porous layer of higher refractive index than both the core and the cladding. The light rays interact with the sensitive intermediate cladding many times, amplifying the spectral modulation more than with a tip coated or evanescent wave FOCs.

- *Refractive index FOCs*
  A bare fibre core, with a thin extra cladding layer of an organic or inorganic compound near the distal end. A change in the refractive index of the analyte surrounding the fibre end alters the transmission properties of the fibre and thus modulates the light reaching the detector. The extra outer cladding must have a refractive index that matches the refractive index of the analyte.

FOCs often use membranes or sol-gels, to contain the reaction chemistry or fluorescent dyes, to protect the chemistry from undesirable species in the environment being monitored, or to allow selective permeation of analytes. Many FOCs use sol-gels as a porous glass tip for silica based fibres. sol-gel provides a solid, inert and porous support rigidly attached to an optical fibre, giving a large, chemically inert surface area [11,20,21,22,23], that enables containment of a wide range of chemically sensitive molecules. Chemicals immobilised in sol-gels for sensing purposes include: fluorescein isothiocyanate dye, which changes its absorption as a function of pH [23], as well as various colourimetric pH indicators, anions, cations, heavy metals, acids, ammonia and enzymes.

Since FOCs were the first optical fibre sensors, based on well-known chemical reactions, they are well developed, widely accepted, and thus are being commercialised [19].

A FOCs has been developed for multicontaminant monitoring of environmental systems such as ground water systems, using an indicator, PRODAN, bound chemically to the cladding on the optical fibre. When excited in the UV, PRODAN fluoresces at various wavelengths, in response to an array of antropogenic organics including acetone, cyclohexane, chloroform, toluene and cyclohexane [28].

6. Optical biosensors

An optical biosensor incorporates a biologically active material, which alters its optical properties, reversibly and selectively in response to the analyte, usually a chemical species. Biosensors form a new and rapidly expanding field of sensors, using isolated enzymes, intact bacterial cells, mammalian and plant tissue, antibodies and bioreceptor proteins. To date, biosensors have generally used physical transduction processes involving charge transfer such as amperometric and potentiometric electrodes, field effect transistors, or solid-state devices such as peizo electric crystals and surface acoustic waves (SAW) devices [30].

Recently, optical biosensors have been developed, in which the biologically active material is a catalyst which is immobilised at the surface of a single fibre, waveguide or fibre bundle, and the detected species is measured by absorbance, fluorescence or chemiluminescence. An analyte consuming reaction is catalysed by the immobilised biocatalyst and a product of this reaction is monitored. The biocatalyst mediates between the analyte of interest and the transducer by converting the analyte to a detectable species. [32,33,34]
Another class of optical biosensors include fibre optic immunosensors (FIS) [35,36], which offer excellent selectivity via anti-body-antigen recognition, enabling measurement of important compounds at trace levels in complex biological samples. FIS for detection of human and environmental exposure to toxic chemical and biological materials, are based on various spectrochemical mechanisms such as absorption, total internal reflection, fluorescence, evanescent field and surface plasmon resonance.

7. Waveguide and integrated optics sensors

Waveguide and integrated optics technology has given rise to a number of clever designs for sensors, including the deposition of sensing layers on waveguides, using a hollow capillary to act as waveguide and sample holder, and devices such as grating couplers adapted for sensing purposes.

Planar waveguides have advantages over optical fibres and are often used for biosensors [37], in applications where the flexibility and remote capability of an optical fibre are not essential.

These advantages include the following :

Total internal reflection fluorescence (TIRF) can be used with lipid membrane structures stabilised onto optical surfaces in planar waveguides.

Planar waveguides allow control over incident reflection angle, enabling control of penetration depth of evanescent field. Langmuir Blodgett deposition of membranes onto planar substrates is well characterised, but difficult onto cylindrical surfaces.

Polarisation information is lost in multimode optical fibres, but retained in planar waveguides. Interferometric measurements possible. Single mode optical fibres would allow polarisation maintenance and interferometry but are not yet available at short wavelengths required for fluorescence in terms of fibre diameter or fibre material. Planar waveguide structures can be placed in flow cell modules.

An optical sensor using inhomogeneous waveguiding in the walls of capillary tubes (capillary waveguide optrodes) [37], also has several advantages over an optical fibre based sensor. The capillary has a chemically sensitive coating on its inner surface which measures analytes in gaseous and liquid samples. Light from a laser or LED is coupled into and out of the capillary, which acts as a waveguide. Light is absorbed in the chemically sensitive coating as a function of the concentration of the analyte.

Advantages include :

- The capillary acts as a sample cavity or flow cell for the gas or liquid and holds a well defined volume.
- Suitable for direct or on-line sampling.
- Optical path length is much longer than other optrode designs and can be optimised for the particular analyte.
- The sensor is not affected by colour and turbidity of the sample.
- Cheap LED light sources and detectors can be used.

Optical devices such as grating couplers are used in integrated optics to couple light into and out of waveguides. Such a device can be used to measure refractive index. The angle at which light couples into a waveguide is a function of the refractive index of the waveguide; if the refractive index changes in response to an ambient medium, then a sensor for refractive index measurement of fluids, humidity or gas concentrations can be constructed. Advantages include small size of sensing head and no moving parts [38].

Optical surface plasmon resonance (SPR) sensors measure the refractive index of a sample near the sensor surface. SPR is an optical effect, in which incident light excites a charge density wave at the interface between a metal and a dielectric. Reflected light decreases in intensity at a certain angle or wavelength, corresponding to a resonant wave vector, whose location depends on the optical characteristics of the medium surrounding the sensor surface. SPR has been used to detect concentrations of 10 ppm Cu and 13.76 ppm Pb [39].

8. Examples of optical sensors for the water industry

Here an attempt is made to match optical sensors and techniques reported in the literature, to the parameters shown in Table 1. Often, sensors have been developed to measure parameters for other applications, such as medical, biotechnology or process control, which can be adapted to the water industry :

- *Multiparameter Water Monitor based on LEDs and colourimetric tests*

  A LED based portable sensor system using standard colourimetric tests uses a multipurpose flow cell with an array of LED sources of different colours, to enable measurement to be made of aluminium, iron, manganese and phosphate levels, as well as sample colour and turbidity [11].

- *Ammonia*

  A fibre optic LED based fluorescence based sensor for ammonium ions uses a pvc membrane containing nonactin, a plasticiser and a proton carrier (lipophilic nile blue). The ammonium ions are extracted into the pvc membrane and protons are released by the dye, contained in the pvc membrane, into the sample solution. Upon deprotonation, the dye undergoes a colour change from red to blue, which is detected fluorescence intensity measurement. A detection limit of 0.03 to 10 mM is achieved with a response times of 1 minute. [40].

- *Nitrates*

  Commercial nitrate analysers use ion specific electrodes, colourimetry and UV absorption. A study of their specifications, as provided by the manufacturer, and their response to waste water and laboratory samples, indicated that the UV based systems were simpler and quicker to use and required less sample conditioning than colourimetric methods. The ion specific electrode analysers were affected by contamination [55].

- *Metal ions and chloride*

  "Optodes" use highly selective ion sensing films to detect ions such as Na^+, K^+, Ca^{2+}, and Cl^-, by optical absorption, fluorescence or reflection mechanisms. In this case, the optical response is generated using ion-pair extraction of the ion-ionophore complex and a deprotonated or protonated lipophilic anionic dye, contained within plasticised PVC membranes. The optodes are designed for disposable use and detect ions by immersion in a glass vessel containing the sample solution, and measurement of the absorption change associated with the lipophilic anionic dyes [59]. Such a sensor is designed for use with serum, but could be adapted for water.

- *Toxicity*

  Microbial toxicity can be measured using a respirometer to measure oxygen consumption. Luminescence quenching by molecular oxygen is the light modulation process, using a luminescent ruthenium complex adsorbed on amorphous fumed silica

and immobilised inside silicone rubber. Luminescent intensity depends on the oxygen concentration of the sample in contact with the sensing film. This sensor was evaluated by measuring the inhibition effect of heavy metals on the respiration rate of micro-organisms in activated sludge, and indicated that IC_{50} (concentration of a toxicant producing 50% respiration inhibition), could be measured [43].
An optical fibre toxicity biosensor, for point or on-line monitoring of toxic effluents, measures the rate of hydrolysis of fluorescein diacetate (FDA) by micro-organisms, which is proportional to their metabolic rate. Rate of increase of fluorescence is proportional to the metabolic rate of the organisms, thus indicating the toxicity of their environment. The organisms are contained within a glass tube, round which is wrapped an electroluminescent film, emitting at the excitation wavelength. Fluorescent light is detected by scattered coupling of light into the walls of the glass tube, in which are embedded an array of fluorescent-core polymer fibres, forming a detecting bundle. The light coupled into the walls of the tube is preferentially detected and amplified by the fluorescent fibre, which is insensitive to the excitation light. The sensor can detect concentrations of fluorescein down to ppb [44].

- *Metals*

Surface plasmon resonance (SPR) techniques can be used to measure metal ions; concentrations down to 500 nM lead and copper[45], and 10ppm copper and 13.76 ppm lead [39] have been measured.

- *Organic solvents*

An optical fibre dipping sensor was developed for detection of organic solvents in waste water, using changes in the optical absorbance of triphenylmethane dyes, measured in the visible part of the spectrum. The sensor layer was separated from the water by a gas permeable membrane. Organic solvents in water were measured down to 30ppm [46].

Chlorinated hydrocarbons such as chloroform and trichloroethylene, can be detected by colourimetry at 560nm, monitored by optical fibres. Pyridine, exposed to chlorinated hydrocarbons, become red, absorbing in the green part of the spectrum. Incandescent lamp and bandpass filter or 555 nm LED and Silicon photodetector was used [47].

- *Suspended solids*

A particle concentration sensor uses two, parallel, plastic 1mm diameter fibre with an extra layer of higher refractive index optical cladding, to make the fibres "leaky". Light in the transmitting fibre, leaks out and undergoes multiple scattering by measurand particles in aqueous suspension, before entering the receiving fibre, which is similarly sensitised. The sensor can be used in point or distributed mode and has been used to measure yeast concentration in the range 0 to 16 g/l. It has been calibrated in turbidity units (FAU) over a range from 0 to 400 FAU, and optimised for the 0 to 20 FAU range. It is presently being optimised for the 0 to 4 FAU range, and also to measure activated sludge concentration [48,49].

- *Cyanide*

A FOCS for cyanide detection, exploits the reaction of metalloporphrins with cyanide. A membrane was used to trap the reagent solution at the common end of a bifurcated fibre-optic bundle.

The cyanide in the sample solution is buffered at a pH low enough to convert to hydrogen cyanide, which diffuses through the membrane and reacts with the reagent. The changes in the optical properties of the reagent can be related to the concentration of cyanide in the sample solution [50].

- *Pesticides*

A immunosensor configured as a waveguide Mach Zehnder Interferometer (MZI), can detect a layer of pesticide of 2×10^{-3} nm, using a well known antibody/antigen system (α-hSA/hSA). The waveguide, contained within a flow cell, uses its evanescent field, associated with guided light within the waveguide, to detect binding processes close to the surface. An immunoreaction at the surface causes a change in the refractive index profile, altering the waveguiding properties of the sensor. The analyte molecules are specified by the immunochemical reaction which takes place at the antibody coated sensor surface [56].

A fibre-optic system based on solid-state, optoelectronic components, including yellow LEDs and photodiodes, detects pesticides because they inhibit the enzyme acetylcholine esterase. A yellow synthetic enzyme substrate is hydrolysed by the enzyme to give a blue product, whose rate of production is reduced by the presence of an inhibitor [57].

- *Fuels and oils*

Near infra-red photoacoustic spectroscopy can be used to detect analytes. In weakly absorbing liquids, the conversion of a pulse of optical energy into acoustic energy, is affected by the composition of the liquid. The properties of the acoustic wave generated enable oil contaminated water to be measured from 0 to 900 mg l^{-1} [58].

- *Dissolved oxygen (DO) and Biological Oxygen Demand (BOD)*

The advantage of optical oxygen sensors over conventional amperometric devices, are that they do not consume oxygen, and are not easily poisoned, small size, no requirement for reference, and inertness against sample flow rates and stirring. Fluorescence quenching of Ruthenium complexes in the presence of oxygen appear to be the sensing mechanism in reported sensors.

One configuration of a DO sensor uses a Ruthenium complex immobilised in a thin film sol-gel, which has strong absorption in the blue-green region of the spectrum, compatible with blue LEDs [11]. A second type of optical oxygen sensor also uses a Ruthenium complex as an oxygen sensitive fluorescent dye, circulated through a loop of oxygen permeable silicone tubing immersed in test medium. The intensity of fluorescence is related to oxygen tension [41]. A third type of sensor for measurement of both dissolved oxygen and BOD, also uses a Ruthenium complex, embedded in a silicone rubber film, which is attached to the inner surface of a transparent vessel in which dissolved oxygen is measured. An optical scanner provides the excitation light and measures the emitted light from the sensing surface and can monitor oxygen in many vessels, simultaneously [42].

- *Total organic carbon*

A comparison of different methods of measuring TOC, examined UV absorption spectroscopy at 254 nm, optical fluorescence with excitation at 340 nm, emission at 420 nm, and established chemical methods of measuring TOC in water samples, together with a commercial UV on-line absorption monitor. The study indicated that such techniques could be adapted for field based instruments using optoelectronic devices. [51]

- *Algae and algal blooms*

Chlorophyll fluorescence of algae can be used to measure presence of algae, these has been performed remotely using satellite imaging systems, as well laboratory based spectrofluorimetry [52]. Time dependent fluorescence techniques, using laser excitation can be used to monitor the effects of pollutants on algae [53].

- *Bacteria and viruses*

Fluorimetric sensors were used to sense CO2 produced by bacterial metabolism, based on fluorescence [54].

- *pH*

Two pH indicators dyes were co-immobilised in a sol-gel film, coated on the core of an optical fibre, so that the pH dependent changes in the cladding could be monitored by evanescent field absorption. Indictors included bromophenol blue, and bromocresol purple. A pH range of 3 to 10 pH units was measured, using a tungsten halogen lamp as the light source, and an Ocean Optics spectrometer for detection [11].
An example of a non-colourimetric indicator pH sensor exploits the surfacic adsorption of positively charged methylene blue dye, chromophores onto the negatively charged surface of a stripped core 200 μm PCS fibre. The adsorbed chromophores interact with the evanescent field and affect the absorption of light guided through the fibre, as the pH of the solution increases. The methylene blue solution is encapsulated within a semipermeable membrane. The linear range of operation is 2 to 10 pH units with a a sensitivity of 0.75dB/pH, and the sensor performs well at low and high ionic strengths, since its sensing mechanism directly exploits OH- ions activity [8].

- *Flow*

An optical fibre based flowmeter, uses two optical fibre strain sensors to measure the strain generated in two orthogonal directions, when placed on adjacent sides of a beam deflected by fluid flow. Strain magnitude is proportional to the square of fluid velocity magnitude. Vectorial addition of the orthogonal strain components give the strain magnitude and direction [60].

9. Commercial optical measurement systems

A survey of commercially available optical sensors [1] indicates that spectrometer and spectrophotometer based sensors, for measuring chemical fluorescence and absorption dominate the market. A selection of commercial optical sensors and measurement techniques measuring parameters useful to the water industry, are briefly described, as follows :

- *Polytec in-line process measurement using a diode array photometer*

Polytec [61] have an X-DAP fibre optic based system, with a series of light sources illuminating samples, and being split into component wavelengths, by a dispersing prism and detected by a diode array. Polytec Procell DAS allows in-line measurement and control of concentrations and colours, continuous monitoring of production process, monitoring of output of waste materials. The complete source, flow cell and detection package is intended for integration into pipes and other parts of process systems, and has no moving parts. The diode array signal is interfaced to a PC.

- *Fibrechem FOCS systems*

Several FOCS systems available, developed by Fibrechem Inc, USA, in collaboration other companies including Texas Instruments (USA) [19]; These included Petrosense CMS (continuously monitoring permanent installation) and Petrosense, Aquasense and Aersense, which are all portable devices. All four systems have the ability for performing on-site, in-situ monitoring of vapour or water samples. Petrosense monitors hydrocarbons, detects leaks smaller than required by legislation (0.2 gal h-1). The lower limit of detection of vapour is 10ppm for xylene petroleum hydrocarbons, and 0.1 ppm for hydrocarbons in water [62]. The Aquasense unit can be used with solid state or reservoir probes, can operate in absorption, reflection or refractive modes and can handle 19 different sensors [19]. Small, low cost sensors, using 4 to 16 individual sensors on a chip on 2.5 mm2 in size, and to house electronics in a second chip of a similar size, are under development. Parameters to be sensed include hydrocarbons, trichloroethylene, carbon dioxide, carbon monoxide, oxygen, ammonia, cyanide, phosphate, sulphate, nitrate and nitrite, trichloroethane, chloroform, carbon tetrachloride and other organics, arsenic, selenium, ferrous iron, manganese, zinc, cadmium, lead, total heavy metals, total iron, mercury and silver.

- *Toxicity Measurement with Eclox, Aquanox and Microtox*

Both the Eclox (Aztec Environmental & Control Ltd) and Aquanox (Randox Laboratories) hand held monitors uses an enhanced chemiluminescent reaction; free radical reaction of the oxidation of luminol in the presence of horse radish peroxidase enzyme, using p-iodophenol as an enhancer and to stabilise the reaction [63, 64] . Any substance, such as an antioxidant pollutant, that inhibits the enzyme reaction, causes a reduction in light emission which is measured on a portable luminometer. The advantage of these units is that they are relatively cheap (Stg 1-4K), portable and can be used by inexperienced operators. Disadvantages include insensitivity to some toxins. Microtox uses a freeze dried marine bacteria; Photobacterium phosphoreum, stored in a cooled storage area within the instrument. In use, a standard amount is rehydrated and mixed with the water sample. The bacteria emit light under favourable conditions and reduce their emission when exposed to toxins. Microtox is widely used in the water industry but it is large, expensive (>Stg 20K) and laboratory based, though an on-line monitor and a hand-held version are under development [65, 66].

- *EauxSys portable turbidity monitor*

A battery powered, portable turbidity meter is designed for low turbidity range of 0.00 to 50.00 NTU. It can be configured for batch or for flow measurement. Light scattered by the particles in aqueous suspension, is measured at 90° to the incident light, giving a true nephlometric measurement [67].

- *Ocean Optics Fibre Optics Sensors for pH, Toxic Metals and Oxygen*

Ocean Optics [68] has a range of optical fibre sensors, coupled to miniature spectrometers to monitor colour changes in immobilised dyes. They form an excellent example of commercialised OFCS, using well-known colourimetric or fluorescent chemical reactions. A tungsten halogen light source and a fibre optic film probe is used with a choice of reactive films. The pH sensitive films are solid state versions of traditional dissolved pH indicators, and each dye has a 3 to 4 pH range, with a precision of ±0.001 pH units. The sensors can be PC interfaced, using custom made Windows compatible software for data acquisition and full spectral analysis. The toxic metal films are immobilised complexing agents, which bind metals in a irreversible reaction, to give an unique spectrum for a particular metal. ppm levels can be detected in 30 minutes.

The oxygen sensor uses a well-known fluorescence quenching reaction to measure oxygen concentration. An optical fibre probe has a proprietary thin film coating on its tip, and a blue LED is the excitation source, with a miniature spectrometer as the detector. Fluorescence detected at the probe is reflected back to the detector, where a 550 nm cut-off filter blocks the excitation light.

- *MetriCor*

Metricor optical fibre sensors use thin film interference as a sensing mechanism. A miniaturised interferometric cavity is placed at the end of a multimode fibre. The optical thickness of the cavity is modified by the measurand, causing a spectral shift. The parameter of most interest to the water industry is refractive index, which can be used to monitor fluids. The optically resonant cavity is exposed to the measurand fluid, causing a change in optical path, and thus a spectral shift that can be measured [69].

10. Conclusion

The range of optical sensors and measurement techniques continue to expand, as new optoelectronic devices and materials become available, driven by the massive optical fibre communications and optoelectronics market. Optical fibre based sensors have been in existence for over two decades and researchers are moving towards novel sensors based on waveguides, integrated optics and new materials.

However, the water industry, driven by legislation and relying on traditional analytical methods, is slow to take up new monitoring technology, and this is indicated by the fact that optical sensors and techniques which have reached the market place, are based on well known colourimetric and fluorescence reactions, measured by spectrophotometric techniques and using optical fibres. Medical, biotechnology and process control industries are adventurous in exploiting new optical technology and many promising sensor principles and techniques are being encouraged by this sector, and will eventually filter through to more conservative industries. The review of new developments and sensors at the research stages indicate that optical techniques can be used to measure all the parameters indicated in Table1, and that optical techniques can offer advantages over existing diagnostic and monitoring techniques. Problems to be overcome by the optical designer include biofouling, calibration stability and selectivity.

REFERENCES

[1] Grattan KTV & Boyle WJO, Optical fibre sensor technology for sensing in the water indusry.CIWEM Advanced Sensors for the Water Industry, Manchester, UK, 1996.
[2] Culshaw B., Optical Fibre Sensing and Signal Processing.Peter Peregrinus Ltd, London, UK, 1984.
[3] Dakin J.P and Culshaw B, Optical Fibre Sensors: Principles and Components.Artech House, Norwood. MA, USA, 1998
[4] Udd E, Fibre Optics Sensors. Wiley Interscience, New York, USA, 1991.
[5] Chapman & Hall, Editors: Grattan KTV & Meggitt BT , Optical Fibre Sensor Technology II-Devices and Applications., London, UK, 1997.
[6] Culshaw B., Optical Systems and Sensors for Measurement, J.Phys E: Sci. Instrum 16, 978-986 (1983).
[7] Koike Y., Staus of POF in Japan, POF'96. Fifth International Conference on Plastic Optical Fibres & Applications, Paris, France, October 22-24,1996.
[8] Deboux B. J-C., Lewis E., Scully P.J and Edwards R, A Novel Technique for Optical Fibre pH Sensing based on Methylene Blue Adsorption, J. Lightwave Technology, 13 (7), 1407-1414, 1995.
[9] Berman E., Driving Bandwidth, POF'96. Fifth International Conference on Plastic Optical Fibres & Applications, Paris, France, October 22-24,1996.
[10] Laguesse M., Sensor applications of fluorescent plastic optical fibres, POF'93 Second Annual Conference, The Hague, June 23-29, 1993
[11] MacCraith B.D., Water monitoring using advanced optoelectronic instrumentation. CIWEM Advanced Sensors for the Water Industry, Manchester, UK, 20 November 1996.
[12] Betlz M.et al , Water quality monitoring using fibre optics at wavelengths below 230 nm. Conference on Optical Science, Engineering & Instrumentation : Chemical, Biochemical & Environmental Fibre Sensors VIII, Denver, August 1996. Proc. SPIE 2836. 1996.
[13] Ball P., Into the blue, New Scientist 2075, 28-31., 1997.
[14] O'Reilly E, Semiconductor lasers go into the blue, Physics World, 9 (4), pp19, 1996.
[15] Marshall L, Biological Monitoring forseen with ultraviolet light source. Laser Focus World April 1994, pp 83.

[16] Butcher S., Optical parametric oscillators open new doors to researcher, Photonics Spectra, May 1994 pp133-138.
[17][Merchant D.F., Scully P.J., Edwards R. and Grabowski J., Optical fibre toxicity bio-sensor. Institute of Physics Applied Optics & Optoelectronics Conference, 16-19 September 1996. University of Reading.
[18] UV photo-induced structures on plastic optical fibres.Schmitt N.F., Lewis E. and Scully P., POF'96, Fifth International Conference on Plastic Optical Fibres &Applications, Paris, France, October 22-24 1996.
[19] Klainer, S.M. et al, Fiber-optic chemical sensors offer a reastic solution to environmental monitoring needs. Sensors & Actuators B, 11 (1993), 81-86.
[20] Bromberg A. et al, Optical fibre sensors for blood gases and pH based on porous glass tips, Sensors & Actuators B31, 181-191.
[21] Grattan KTV et al, Sol-gels for fibre optic sensors, Proc.6th Int. Optical Fibre Sensors Conference, Paris, Arditty H.J., Dakin J.P., Kersten R. Th (Eds), Optical Fibre Sensors (Springer Verlag, Berlin) pp 436-42, 1989.
[22] Klein L.C (Editor), Sol-gel Optics : Processing and Applications. Kluwer Academic Press, 1994.
[23] Badini et al, Characteristics of dye-impregnated tetrathylorthosilane (TEOS) derived sol-gel coatings. J. Sol-Gel Sci & Tech. 6, 262-72, 1996.
[24] Klainer S.M. et al, Environmental monitoring applications of fiber optic chemical sensors (FOCS), Fiber Optic Chemical Sensors and Biosensors Volume II, Editor : Otto S. Wolfbeis. CRC Press, 1991.
[25] Bogue R.W(1995). Environmental Sensors : Present practice and future prospects, Sensors and their Applications VII : Proceedings of the seventh conference on Sensors and their Applications, Dublin, Ireland, 10-13 September 1995, Editor: A.T. Augousti. Institute of Physics Publishing, Bristol. UK.
[26] Bogue R.W, Sensors of the Future. CIWEM Sensors & Sensitivity : Water quality control through instrumentation and sensors. City Conference Centre, London. 6 December 1995.
[27] Burgess L.W., Absorption-based sensors. Sensors and Actuators B29, 10-15, 1995.
[28] Leonard K.L, Development of a fibre-optic chemical sensor for multicontaminant monitoring of environmental systems. Sensors & Actuators B 24-25, 458-461, 1995.
[29] Hale Z.M. et al, The single mode tapered optical fibre loop immunosensor, Biosensors & Bioelectronics 11, (1/2), 137-148, 1996.
[30] Arnold M.A and Wangsa J., Transducer based and intrinsic biosensors. Fiber Optic Chemical Sensors and Biosensors Volume II. Editor: Otto S. Wolfbeis. CRC Press, 1991.
[31] Brecht A. & Gauglitz G, Optical probes and transducers, Biosensors & Bioelectronics, 10, 923-936, 1995.
[32] Robinson G., The commercial development of planar optical biosensors, Sensors & Actuators B 29, 31-36, 1995.
[33] Kobatake E. et al, Biosensing of benzene derivatives in the environment by luminescent escherichia coli. Biosensors & Bioelectronics 10, 601-605, 1995.
[34] Brecht A. & Gauglitz G, Optical probes and transducers. Biosensors & Bioelectronics 10, 923-936, 1995.
[35] Vo-Dinh T. et al, Fibreoptics Immunosensors. Fiber Optic Chemical Sensors and Biosensors Volume II. Editor: Otto S. Wolfbeis. CRC Press, 1991.

[36] Plowman T.E. et al, Femtomolar sensitivity using a channel-etched thin film waveguide fluoroimmunosensor, Biosensors & Bioelectronics, 11 (1/2), 149-160, 1996.
[37] Weigl B. H., Lehmann H. & Lippitsch M.E, Optical sensors based on inhomogeneous waveguiding in the walls of capillaries, Sensors & Actuators B32, 175-179, 1996.
[38] Brandenburg A. & Gombert A, Grating couplers as chemical sensors; a new optical configuration. Sensors & Actuators B, 17, 35-40, 1993.
[39] Jung C.C. et al, Chemical electrode surface plasmon resonance sensor, Sensors & Actuators. B32, 143-147, 1996.
[40]Wolfbeis O. & Li Hong, Led compatable fluorosensor for Ammonium ion and its application to biosensing, SPIE Vol 1587, Chemical, Biological and Environmental Fiber Sensors III, pp48-58, 1991.
[41] Towe B. et al, A recirculating-flow fluorescent oxygen sensor, Biosensors and Bioelectronics. 11, (8), 799-803, 1996.
[42] Li X.M. et al, Scanning optical sensor for the measurement of dissolved oxygen and BOD, Sensors & Actuators B 21, 143-149, 1994.
[43] Wong K.Y. et al, A luminescence based scanning respirometer for heavy metal toxicity monitoring, Biosensors & Bioelectronics 12, (20), pp125-133., 1997.
[44] Merchant D., Scully P.J., Edwards R. and Grabowski J., Optical fibre fluorescence and toxicity sensor. Environsense '97, European Symposium on Environmental Sensing III 16-20 June 1997, Munich Fairgrounds, FR Germany, 1997.
[45] Chirowsky T. et al, Experimental data from a trace metal sensor combining surface plasmon resonance with anodic stripping voltametry, Sensors & Actuators B35-36, pp 37-43, 1996.
[46] Dickert F.L et al, Fiber-optic dipping sensor for organic solvents in waste water, Anal.Chem. 61, 2306-2309, 1989.
[47]Milanovich F.P et al, A fiber-optic sensor for monitoring chlorinated hydrocarbon pollutants., Talanta 41, (12), pp2189-2194, 1994.
[48] Zhang F.H., Scully P.J, and Lewis E., A novel optical fibre sensor for turbidity measurement, Institute of Physics Applied Optics & Optoelectronics Conference, 16-19 September 1996, University of Reading, 1996.
[49] Zhang F., Scully P.J. and Lewis E., An Optical Fibre Sensor for On-Line Yeast Measurement, 1994, Institute of Physics: Applied Optics and Optoelectronics Conference. University of York, Oral Presentation, September 1994.
[50] Freeman M.K. & Bachas L.G., Fiber-optic probes for cyanide using metalloporphyrins and a corrin, Analytica Chimica Acta, 241, 119-125, 1990.
[51] MacRaith B. et al, Cross comparison of techniques for the monitoring of total organic carbon (TOC) in water sources and supplies., Wat. Sci. Tech. 28, 11-21, pp457-463, 1993.
[52] Gitelson A. and Yacobi Y., Spectral features of reflectance and algorithm development for remote sensing of chlorophyll in Lake Kinneret, Air Toxics and Water Monitoring. Europto, SPIE Vol.2503, 21 June 1995.
[53] Ruth B., Monitoring the herbicide concentration in ponds by detecting the chlorophyll fluorescence of algae with ms-laser-excitation, Air Toxics and Water Monitoring, Europto 21 June 1995, SPIE Vol.2503, 1995.
[54] Swenson F.J., Development and evaluation of optical sensors for the detection of bacteria, Sensors & Actuators, B11, 315-321, 1993.

[55] Wacheux H. et al, Inventory and assessment of automatic nitrate analysers for urban sewage works, Wat. Sci. Tech. 28, (11-12), pp489-498, 1993.

[56] Lechuga L.M. et al, Feasibility of evanescent wave interferometer immunosensors for pesticide detection : chemical aspects, Sensors & Actuators B 24-25, pp762-765, 1995.

[57] Trettnak W. et al, Fiber-optic remote detection of pesticides and related inhibitors of the esterase : acetylcholine esterase, Sensors & Actuators B11, 87-93, 1993.

[58] Hodgson P. et al, Application of pulsed laser photoacoustic sensors in monitoring oil contamination in water, Sensors & Actuators B29, 339-344, 1995.

[59] Hisamoto H. et al, Ion sensing film optodes : disposable ion sensing probes for the determination of NA+,K+,CA2+, and Cl- concentrations in serum, Sensors & Actuators B 29, 378-385, 1995.

[60] Chandy R., Scully P.J. and Morgan R., Elastic beam deflection flowmeter using optical and conventional strain gauges, Institute of Physics Applied Optics & Optoelectronics Conference, University of Reading, 16-19 September 1996.

[61] Lambda News Spring, Lambda Photometrics Ltd. UK, 1997.

[62] Saini DPP, FCI Environmental Inc, Publicity Data, 1996.

[63] Aquanox publicity data, Randox Laboratory Ltd. UK.

[64] Hayes E. & Smith M., Eclox : a rapid screening toxicity test. International Conference on Toxic Impacts of Waste on the Aquatic Environment, Loughborough University, UK, April 1996.

[65] Bartlett-Hooker B., On-line toxicity monitoring. International Conference on Toxic Impacts of Waste on the Aquatic Environment. Loughborough University, UK, April 1996.

[66] Microtox publicity data, Microbics (UK) Ltd.

[67] Type 2300P Portable Turbidity Monitor publicity data, EauxSys (UK) Ltd.

[68] Ocean Optics Europe publicity data, The Netherlands.

[69] Metricor Sensors, Analusis Magazine, 22, (3), 1994.

MICROBIOLOGICAL SENSORS FOR THE MONITORING OF WATER QUALITY

D. OSBILD and P. VASSEUR, Centre des Sciences de l'Environnement. Université de Metz., 57 000 Metz, France

1. Introduction

The environmental monitoring of pollutants with automatic systems, applied on-line and allowing rapid responses constitutes one of the most successful ways to improve the quality of the environment. Real time toxicity analysis offers the advantage of rapidly detecting sources of pollution and preventing any accidental release of pollutants. There is a need for in situ continuous biomonitoring of underground waters, surface waters and rivers, of pollution sources, and of the efficiency of depuration processes.. Such a strategy is not possible with conventional toxicity tests, carried out in the laboratory and in static conditions. In situ continuous biomonitoring has thus encouraged the development of biological alarm systems, giving a rapid response, in real time, which can be automatically recorded.

In order to fulfil these requirements, the concept of these alarm systems must differ from classical aquatic toxicity tests :

- their response must be delivered after a short period of exposure of the biocatalyst to the flow : this implies following other relevant criteria than growth, mortality or reproduction,
- the biological response must be easily recorded and converted into an electric signal which may set off an alarm. A suitable association of the biocatalyst and the transducer is needed to ensure the performance of the system,
- an immobilization is necessary when using whole cells as biocatalysts, as they are used in flow-through conditions.

Environmental biomonitoring and toxicity assessment require devices with a broad spectrum sensitivity rather than highly specific ones. As the array of contaminants is wide and the threat unknown, the choice of cellular rather than molecular systems is more suitable.

If there is a need to increase testing for an improved protection of the environment, there is also a need for fast, simple and cheap ecotoxicological methods. Microbial biosensors probably show a greater potential for pollution monitoring and toxicity assessment than other alarm warning systems using aquatic invertebrates and vertebrates, or immobilized enzymes and antibodies.

2. Microbial biosensors

Whole cell biosensors probably offer the greatest technological changes among early warning systems, with an immobilized biocatalyst (in most cases) in close contact with a transducer unit. Their development has benefitted from biotechnological progress in enzyme electrodes used to control fermentation processes in food technology and medical analyses. To date, bacteria, yeasts and photosynthetic cells (cyanobacteria and microalgae) constitute the main biocatalysts used in microbiological biosensors.
Whilst some microbial biomonitors propose the use of free cells in bioreactors, most systems utilize biocatalysts immobilized on membranes or entrapped in gel matrices.

Microbiological biosensors combine the advantages of the two other categories of early warning systems (E.W.S.) developed today for aquatic environmental biomonitoring : in one hand, enzyme and immunological electrodes, and E.W.S. using vertebrates and invertebrates in the other hand. Microbiological biosensors share the advantages of molecular electrodes :

- the miniaturization of systems using immobilized cells,
- the immobilization, which ensures practical handling in flow-through conditions.

Cells must be immobilized by gentle processes in order to maintain cellular integrity : these processes are restricted to adsorption or entrapment in non-toxic polymers. The weakness of adsorption can lead to leakage problems of the biocatalysts in the flow. On the other hand, gel matrix is permeable to organic substances but impermeable to cations, the toxicity of which will not be detected by entrapped cells.
In contrast to molecular electrodes which exploit only one combination enzyme/substrate or antigen/antibody, microbial biocatalysts are living cells, i.e. complete organisms with multiple biochemical pathways governed by multiplicity of enzymes, which thus offer the greatest potential of investigation. Therefore, microbial biosensors share the property of presenting a wide spectrum of response to toxicants, with vertebrates and invertebrates : this has already been underlined as an incomparable advantage in environmental biomonitoring. Yet, it must not be forgotten that microbial biocatalysts are alive and must be kept alive to remain active. Consequently, nutrients must be provided during the servicing time of the biosensor : this is a source of problems in the conception and the construction of the device and in its sensitivity.

All these systems must deliver a signal after a short period of exposure to toxicants : this influences the choice of the parameter(s) to be monitored because biological changes in the short term must be representative and predictive of toxicity in the long term. The biomonitoring of energetic cellular processes is privileged because of their relevance as criteria for toxicity. Indeed, ATP levels control a number of essential physiological events, such as motility, synthesis of cell constituents, growth, division, viability,....and any impairment of ATP synthesis will have dramatic consequences on these processes. Respiratory activity and photosynthesis are the main metabolic pathways monitored in microbial biosensors (tables 1-4). Respiration can be easily followed by measuring oxygen consumption or CO_2 production with an electrochemical transducer ; photosynthetic activity will also be controlled by measuring O_2 production after exposing the cells to light.

Optrodes are transducers of interest due to their high sensitivity for measuring the fluorescence of photosynthetic pigments or bioluminescence produced during respiration in naturally luminescent bacterial cells (tables 2 and 3).

Measuring electron transfer between redox components of the respiratory or photosynthetic pathway has also been recommended. Yet, the transfer of electrons to the electrode often requires the use of mediators, such as quinone derivatives or ferricyanide. Mediators may be toxic to the biocatalyst in the long term, which will impair its longevity.

New systems using genetically transformed microorganisms are now emerging : these systems exploit the simplicity of luminescent measurements in order to identify the activation of genes as a result of exposure to pollutants. Lux gene is inserted in the promoter region of the genes controlled, whose transcription will be expressed by light emission. Thus these systems detect very early events ; their specificity will depend on the role of the genes controlled, whether they express an exposure to a specific toxicant or not. The activation of genes which code for stress proteins for instance, can be considered as a general response. In other cases, the response can be very specific to a particular toxic substance (table 2).

3. Field of application of microbiological biosensors

The degree of sensitivity and specificity of biosensors condition their field of application. A response to a wide spectrum of toxicants is necessary when controlling surface waters, underground waters and wastewaters, whose contamination sources are multiple and pollutants unknown. In contrast, biosensors of high specificity will be devoted to situations with identified polluting substances (Osbild et al., 1995).

Non specific bacterial biosensors generally detect chemicals at concentrations from 10^2 - 10^3 µg/l or µM (tables 1 and 2). This level of sensitivity is a little lower than other E.W.S. using fish or bivalves, and apparently far below the sensitivity of algal biosensors with fluorimetric transducers (table 3).

The field of application of bacterial biosensors as alarm systems is the following :
the detection of high pollution loads likely to induce toxicity in the short term ; this includes the monitoring of rivers used for the production of drinking water,
the survey of complex effluents and wastewaters before their discharge into surface waters,
the monitoring of treatment plants, performed (i) to protect biofilms from an input of high toxic charges, and (ii) to control the efficiency of depuration processes.

The performance of algal biosensors appears better than non-specific bacterial sensors. Nevertheless, most experiments have been carried out with herbicides which are specifically toxic to photosynthetic activity : it cannot be excluded that their sensitivity to other kinds of contaminants may be lower. Nonetheless, photosynthetic cells are the ideal biocatalysts to detect pollutants toxic to the vegetal world. Consequently, they must necessarily be included in a strategy for water biomonitoring.

The sensitivity of microbiological biosensors depends on three main parameters, that is to say :

- the sensitivity of the biocatalysts themselves,
- the nature of the transducer connected to the biocatalyst. For instance, increased sensitivity over two orders of magnitude was seen in algal biosensors when substituting fluorescent receptors to O_2 electrodes (Schmitz et al. 1993 ; Pandard and Vasseur, 1992).
- the environment of the biosensor, which influences the bioavailability of pollutants and thus their toxicity. Physico-chemical composition of the analysed medium, which may contain complexants in addition to suspended solids likely to adsorb pollutants, pH, temperature, oxygenation,... are factors which may interfere in the response of the biocatalyst and need to be taken into account or adjusted (Osbild, 1997).

It is worthy remembering that microorganisms are live cells, which need to be fed during their use. Conversely, high concentrations of nutrients in the flow may decrease the sensitivity in some cases, by maintaining an over active metabolism, or by modifying the bioavailability of contaminants. Therefore, a compromise has to be found in concentrations ensuring simultaneously longevity and stable sensitivity.

Table 1 : Characteristics of the main biosensors using bacteria as biocatalysts and measuring respiration changes

PARAMETRE OR RESPONSE SPECIES	BACTERIA	SUBSTANCE(S)	LOEC* µM	LOEC* µg/l	RESPONSE ....TIME	REFERENCES
O_2						
Methylomonas flagellata	Free	Methane	5		1 min	Karube *et al.*, 1982
	Immobilized Nylon membrane	BOD		2 to 22.10³	10 min	Strand and Carlson, 1984
Bacillus subtilis	Immobilized PVA gel	BOD		2 to 22.10³	30 s	Riedel *et al.*, 1988
Activated sludge "RODTOX"	Free 10 l.	3,5-dichlorophenol Cu^{2+} CN^-		1.10⁴ (IC_{50}) 11.10³ 780	5 min 15 min 5 min	Kong *et al.*, 1993
Pseudomonas putida "Toxalarm"	Free	PCP		10³	Immediate	Schmitz *et al.*, 1993
Pseudomonas putida "Stiptox - norm"	Free			1.5.10³		
Activated sludge "Toxiguard"	Biofilm			50		
Activated sludge "Biox-1000T"	Fixed on rings			10³		
Nitrosomonas eurapae	Immobilized Cellulose acetate membrane	NH^{4+} Arsenic Pb^{2+} CN^- ortho-chlorophenol Trichloroethylene Tetrachloroethylene		0 to 10³ (IC_5) > 50 (IC_5) > 100 (IC_5) 40 (IC_5) 60 (IC_5) 60 (IC_5) 40 (IC_5)	10 min	Tanaka *et al.*, 1993
Escherichia coli	Immobilized	PCP Tributyric tin chloride Formaldehyde Phenol 2-chlorophenol MCPA Mercury chloride Atrazine Cadmium Copper chloride		9 11 5.10³ >10³ 50 150 27 >10³ >10³ 500		Hatton *et al.*, 1994
Bacillus subtilis and *Bacillus licheniformis*	Immobilized Polycarbonate membrane	BOD		0 to 7.10⁴	5 min	Li *et al.*, 1994
LAS degrading strain	Immobilized Calcium alginate gel	LAS (Alkylbenzene linear sulphonate)		0 to 6.10³	15 min	Nomura *et al.*, 1994
Activated sludge	Free	3,5-DCP		5.5.10³ (IC_{50})	30 min	Rawson and Rogerson, 1994
Pseudomonas putida	Immobilized cellulose acetate membrane	Benzene	71 to 11.10³		2-10 min	Tan *et al.*, 1994
Activated sludge "Biox-1000T"	Fixed on rings	BOD		5.10³	3 min	Teutscher and Grosser, 1994

Activated sludge "RODTOX"	Free 10 l.	BOD PCP Hg^{2+} Cu^{2+} CN^- 3,5-DCP o-cresol Toluene	10^4 to 5.10^8 10^3 10^4 10^4 10^4 10^4 10^5 10^6	30 min 1 h	Vanrolleghem *et al.*, 1994
Thiobacillus thioparus	Immobilized Nitrate cellulose membrane	Thiosulphate Methanethiol	1 to 10^3 200 to 3.10^3	1 min	Kubo *et al.*, 1995
Magnetic activated sludge	Immobilized	BOD	0 to 60.10^3	30 min	Sakai *et al.*, 1995
Sphingomonas species *Pseudomonas fluorescens*	Immobilized Polymethane gel	Naphtalene	30 to 2.10^3	2 min	König *et al.*, 1996
Activated sludge "Rodtox"	Free	Creoline	5.10^3 (IC_{20})	1 h	Kong *et al.*, 1996
Pseudomonas fluorescens	Immobilized Cellulose nitrate membrane	HCN	10^2 to 10^3	3 min	Lee and Karube, 1996
Pseudomonas putida	Immobilized PVA gel	Paraoxon	1 to 10^3	20 min	Rainina *et al.*, 1996
Bacillus stearothermophilus	Immobilized Gel	Phenol o-cresol m-cresol p-cresol 3-chlorophenol Catechol 4-methylcatechol 2-naphtol	2.5 to 50 1 to 3 1 to 3 1to 3 2,5 to 25 5 to 200 5 to 200 2.5 to 70	2 min	Rella *et al.*, 1996
Brevibacterium species	Free	Acrylamide Acrylic acid	10^4 10^4	15 min	Ignatov *et al.*, 1997
Activated sludge (Optrode)	Free	Hg^{2+} Cu^{2+} Pb^{2+} Ni^{2+} Zn^{2+} Cr^{3+}	6.10^3 (IC_{50}) 8.10^3 (IC_{50}) 19.10^3 (IC_{50}) 21.10^3 (IC_{50}) 28.10^3 (IC_{50}) 40.10^3 (IC_{50})		Wong *et al.*, 1997
CO_2 *Escherichia coli*	Immobilized Nucleopore membrane	Phenol HCN Cd^{2+} AsO_2^- Pb^{2+} Cu^{2+}	1.10^6 (IC_{40}) $3.5.10^3$ (IC_{40}) 300 (IC_{40}) $1.5.10^3$ (IC_{40}) 130 (IC_{40}) 1.10^3 (IC_{40})	5 to 30 min	Dorward and Barisas, 1984
Hyphomicrobium	Immobilized	Methylsulphate	10^3	30 min	Schär and Ghisalba, 1985
Pseudomonas aeruginosa	Immobilized Nylon membrane	Nitrate	2.10^4 to 2.10^5	15 min	Hikuma *et al.*, 1993
electronic transport *Escherichia coli* "Bucyano-Bacteria Electrode"		PCP	10^3		Schmitz *et al.*, 1993
Escherichia coli E *Pseudomonas putida* P		PCP PCP	500 E 10^3 P		Hansen, 1994

* Lowest concentration inducing a response (LOEC). The concentrations of inorganic elements are expressed as μg/l or μM of the corresponding cation or anion

Table 2 : Characteristics of the main biosensors using luminescent bacteria as biocatalysts. Comparaison with IC 50, for the same species.

RESPONSE SPECIES	BACTERIA	SUBSTANCES	LOEC* µM	LOEC* µg/l	RESPONSE TIME	IC50 Microtox µg/l 15 min - *5 min	REFERENCES
Vibrio fischeri **Lux CDABE / in :** . *Staphylococcus aureus* S . *Escherichia coli* E	Free	Bismuth Cd^{2+} Pb^{2+}	1 E 2 E 3 S 1 E 10 S		120 min S 60 min E		Corbisier *et al.*, 1993
Escherichia coli	Free	Hg^{2+}	5.10^{-4} to 10^{-3}		40 min to 100 min		Selifonova *et al.*, 1993
Escherichia coli	Free	$HgCl_2$		2 to 800	60 min		Tescione and Belfort, 1993
Pseudomonas fluorescens	Immobilized Glycerol alginate gel	Naphtalene Salicylate		550 500	21 min 24 min		Heitzer *et al.*, 1994
Escherichia coli stress proteins	Free Microplates	Cu^{2+} 2,4-dichloro phenoxyacetic acid Methanol 2-nitrophenol 4-nitrophenol Phenol Xylene		9.10^4 2.10^5 2.10^5 1.10^4 4.10^3 2.10^4 3.10^5	10 min	$*2.10^6$ (a) $*13.10^3$ (a) $*34.10^3$ (a) $*16.10^3$ (a)	Van Dyk *et al.*, 1994
Vibrio fischeri "Automicrotox"	Free Freeze-dried	Na Laurylsulphate Toluene Chlorobenzene Manebe Phenol CN^- Zn^{2+} Cu^{2+} Cd^{2+} Cr^{6+} Pb^{2+}		2.10^3 (IC50) 2.10^4 (IC50) $7.5.10^3$ (IC50) 23 (IC50) 2.10^4 (IC50) 2.10^3 (IC50) 580 (IC50) 730 (IC50) 7.10^3 (IC50) 2.10^4 (IC50) 900 (IC50)	15 min 15 min 15 min 15 min 15 min 15 min 15 min 15 min 15 min 30 min 15 min total duration of a test : 50 min	$1.5.10^3$ 15.10^3 8.10^3 39 $18.5.10^3$ 1.10^3 1 1 7 11.5 1	Heuriet *et al.*, 1990
	Immobilized Cellulose nitrate membrane	Benzalkonium chloride (biocide) Cr^{6+}	23 (IC50) 85 (IC50)		few minutes		Lee *et al.*, 1992
"Photox" = RBT	Free Freeze-dried	PCP		50	15 min		Schmitz *et al.*, 1993
"Biotoximeter"	Free batch			$1.5.10^3$			
"Automicrotox"	Free Freeze-dried	Malathion Chromium Formaldehyde Diesel		50 (IC39) 10^5 (IC51) 65 (IC74) 750 (IC51)		$*12.10^3$ (a)	Chevillon, 1994
Firefly luciferase in *Escherichia coli*	Free	Chloramphenicol Sodium azide Fluoroacetic acid Neomycine sulphate		10^4 (IC12) 50 (IC20) 10^4 (IC12) 10^4 (IC12)	30 min		Lee *et al.*, 1991
Photinus pyralis **Luc FF in** *Escherichia coli*	Free Microplates	$HgCl_2$ $CdCl_2$	10^{-9} to 0.1 0.5 to 100		60 min		Virta *et al.*, 1995

* Lowest concentration inducing a response (LOEC). The concentrations of inorganic elements are expressed as µg/l or µM of the corresponding cation or anion.
(a) Kaiser and Ribo (1988)

RESPONSE	MICROALGAE	SUBSTANCE(S)	LOEC µM	LOEC µg/l	RESPONSE TIME	REFERENCES
O_2 production	*Chlorella vulgaris*	Phosphates	8.10^3 to 7.10^4		1 min	Matsunaga *et al.*, 1984 a
		Isoproturon		100 (IC50)	30 min	Pandard and Vasseur, 1992 Pandard *et al.*, 1993
		Propanil		500 (IC50)	30 min	
		Hg^{2+}		$1.5.10^3$ (IC50)	180 min	
		CN^-		$> 3.10^3$ (IC50)	30 min	
		Chlortoluron		100 (IC60)	30 min	
		Atrazine		100 (IC27)	30 min	
		Cu^{2+}		$1.5.10^3$ (IC70)	420 min	
	Synechococcus species	Atrazine		< 10	10 min	Hansen *et al.*, 1989
		Linuron		100		
		Diuron		200		
		Metoxuron		50		
		Lindane		100		
	Rhodococcus species	Phenol	4		5 min	Riedel *et al.*, 1993
		3-chlorophenol	4			
		3-chlorophenol	4			
		4-chlorophenol	4			
	Chloroplastes and thylakoïdes	Atrazine	1			Purcell *et al.*, 1990
		Diuron	0.07			
		Atrazine		12	5 min	Rouillon *et al.*, 1995 b
Fluorescence and O_2	*Chlorella reinhardii* "Fluox-algae"	Atrazine		1		Schmitz *et al.*, 1993
Fluorescence	*Chlorella vulgaris*	Simazine		12	5 min	Weston and Robinson, 1991
	Chlorella reinhardii "Biosens-algae-toximeter"	Atrazine		2		Schmitz *et al.*, 1993
	Chlorella reinhardii "IFW-fluorometer"			5	5 min	
	Scenedesmus subspicatus "Fluotox"			1	5 min	
				10	60 min	
Delayed fluorescence	*Chlorella reinhardii* "DF-algae-toximeter"	Atrazine		0.4		Schmitz *et al.*, 1993
	Synechococcus	DCMU		223	10 min	Rawson *et al.*, 1987
		Chlortoluron		2.10^3		
		Linuron		17		
		Atrazine		33		Schmitz *et al.*, 1993
		Atrazine		467 (IC50)	5 à 10 min	Van Hoof *et al.*, 1992
		Linuron		626 (IC50)		

Table 3 : Characteristics of the main biosensors using photosynthetic organisms as biocatalysts (part 2).

Electronic transport		Atrazine	10	2 min	Hansen, 1994
		Chlortoluron	10		
		Metamitron	200		
		Diuron	2.5		
		Linuron	50		
		Terbutylatrazine	25		
		Simazine	10		
		Crimidine	25		
		Isoproturon	25		
	thylakoïdes	Diuron	0.45 (IC50)		Carpentier *et al.*, 1989
		Atrazine	7 (IC50)		
		$NaNO_2$	550 (IC50)	3 à 5 min	Purcell et Carpentier, 1990
		$NaSO_3$	2.10^3 (IC50)		
		$CuCl_2$	720 (IC50)		
		$PbCl_2$	10^3 (IC50)		
		$HgCl_2$	2.10^3 (IC50)		
		$ZnCl_2$	2.10^3 (IC50)		
		$CdCl_2$	4.10^3 (IC50)		
		Diuron	2.10^{-2}	5 min	Rouillon *et al.*, 1995 c
		Atrazine	0.2		
		$HgCl_2$	10^3	5 min	Rouillon *et al.*, 1995 a
		$PbCl_2$	10^3		
		$CdCl_2$	10^3		
		$NiCl_2$	10^2		
		$ZnCl_2$	10^2		
		$CuCl_2$	10^2		
Specific electrodes	H_2 *Chromatium species*	S^{2-}	400	5-10 min	Matsunaga *et al.*, 1984 b
	chloride and bromide *Rhodococcus species*	2-Ethylenebromide	40	5 min	Peter *et al.*, 1996
		1-Chlorobutane	220		

* Lowest concentration inducing a response (LOEC). The concentrations of inorganic elements are expressed as µg/l or µM of the corresponding cation or anion.

Table 4 : Characteristics of the main biosensors using yeast as biocatalysts.

RESPONSE / SPECIES	YEAST	SUBSTANCES	LOEC* µM	LOEC* µg/l	RESPONSE TIME	REFERENCES
O_2						
Trichosporon cutaneum	Immobilized PVA gel	BOD		4.10^3 to 10^5	30 s	Riedel *et al.*, 1988
Trichosporon beigelii (cutaneum)	Immobilized PVA gel	Phenol	2		5 min	Riedel *et al.*, 1995
		2-chlorophenol	2			
		3-chlorophenol	2			
		4-chlorophenol	2			
		2,3-dichlorophenol	2			
		2,4-dichlorophenol	2			
		2,5-dichlorophenol	2			
Saccharomyces cerevisae	Immobilized "Chitopearl"	Cyanide	0 to 15		10 min	Ikebukoro *et al.*, 1996
Saccharomyces cerevisae	Immobilized Nitrate cellulose membrane	CN^-	0.3 to 150		2 min	Nakanishi *et al.*, 1996

* Lowest concentration inducing a response (LOEC). The concentrations of inorganic elements are expressed as µg/l or µM of the corresponding cation or anion.

4. Requirements for microbiological biosensors used in biomonitoring

As underlined above, specificity and sensitivity are important to determine the field of application of a biosensor and its use. These two properties are the main characteristics studied to judge the interest of a biocatalyst and its associated transducer. These studies are conducted on laboratory pilots, and generally over a short period of time.

A large number of microbiological biosensors have been proposed on the basis of such results. Yet, this first step must be followed by in situ studies, allowing other criteria to be evaluated :

- longevity,
- storage of the biocatalyst,
- reliability,
- practicability,
- servicing time,
- computer equipment,
- absence of interference leading to false alarms,...

These criteria are crucial in determining the interest of the instrument. Unfortunately, most of the investigations on microbial biosensors are still performed at a laboratory level. The validation step in the open environment has yet to be carried out.

A study initiated by several German institutions was carried out on the River Rhine, from 1990 to 1992, in order to test and compare automated biomonitors on the basis of the above criteria. This study is exemplary of what needs to be done to optimize the implementation of automated biotests. Results highlighted the interest of algal biosensors with fluorimetric transducers (Schmitz et al., 1993).

5. Needs in biosensor technology development

The in situ validation is a necessary step in the development of biosensors. This step takes time, needs much work and is also costly. The difficulty of the operation is expressed by the discrepancy between the large number of biosensors described after feasibility studies in the lab, and the rarity of automated biological systems in use in biomonitoring at the moment. This problem is not restricted to microbial sensors, but general to all automated biomonitors.

In fact, what we call a biosensor is not simply an efficient association between biocatalyst and transducer, but a device with systems which must avoid failure and interference, and including : thermostatisation, addition of nutritive solutions, adjustment of pH and perhaps salinity, exposure of the photosynthetic catalysts to light, elimination of suspended solids if necessary,... All the parameters known to influence the response of the biological reagent need to be controlled. The efficiency of these controls will insure the reliability of the alarm system.

Other hydraulics-related problems may also arise in the device. From our experience, dynamics problems are frequent and often take a long time to be identified because they are generally unexpected. The biocatalyst is always incriminated in the case of failure of the systems : yet, most of the time, problems arise from other causes : mechanics, physics, dynamics.

This enumeration of traps which obstruct the implementation of alarm systems, aims to underline the fact that we should concentrate on studying the devices under field conditions. We have now to focus our efforts on the validation of biosensors in field situations. This is a priority if there is to be a biomonitoring of waters for an improved protection of the aquatic environment. Otherwise, biosensors will remain attractive tools for scientists.

Another area which deserves to be explored is the search for non specific microbial biosensors with a higher sensitivity than the present systems. Lower detection thresholds of toxicants will allow applications which biomonitor the quality of rivers, surface waters and underground waters, characterized by a low level of pollution, but with possible long term toxic effects. At the moment, only specific microbial sensors present such a level of sensitivity. The association of multiple specific biosensors could extend their array of detection. This kind of system would certainly deserve to be implemented. However, an in situ validation will still have to be carried out in order to guarantee the system's reliability for environmental biomonitoring.

The systems, which are fast, simple and cheap, are likely to have the most commercial success and the most wide-spread use in environmental biomonitoring, provided their reliability has been proved in the environment. This enthusiastic goal cannot be reached without the cooperation of biologists, engineers, statisticians, electronicians, this close collaboration being absolutely necessary to ensure success. But above all, the sponsoring of European and governmental institutions is needed, as well as that of industries and commercial societies interested in the development of this new technology. Success will require a team effort, but the interests at stake are great, both environmentally and commercially.

6. Acknowledgements

The authors would like to thank ADEME, the Group Inter-Agences (French Water Authorities) and SERES for their support. They are grateful to Tracy Carmona for her assistance in the preparation of this paper.

DATA MINING OF WATER QUALITY DATA BY CHEMOMETRICAL METHODS

B.G.M. VANDEGINSTE, Unilever Research Laboratorium Vlaardingen, P.O. Box 114, 3130 AC Vlaardingen, The Netherlands.

1. Setting the scene

In the field of water quality monitoring, increasingly complex sets of data are measured, which require advanced data processing technologies in order to extract the information hidden in the data.
In increasing order of complexity we may think of time dependent concentration levels of several constituents at one location or at several locations. This respectively results in a table or matrix of data (time * concentration) or a cube of data (time * position * concentration). A complication specifically with environmental data is that some constituents are not monitored with the same frequency, nor at the same time.

Complexity may even be larger if, instead of concentrations, we consider the output of e.g. a sensor array. If such a sensor array is positioned at several locations in a process, the operator receives consecutive frames of data of the dimension (time * number of sensors * number of locations). A quick judgement of the data in terms of control actions is then required. If these data are stored in a database, the mining of such historical data may be of paramount importance for process modelling.

Obviously, visualisation and modelling techniques should assist the operator in interpreting and modelling the data (Fig. 1). Methods which found a widespread application in environmetrics are pattern recognition, class modelling and factor analysis [1].

Fig. 1 Issues in environmental assessment (adapted from [1])

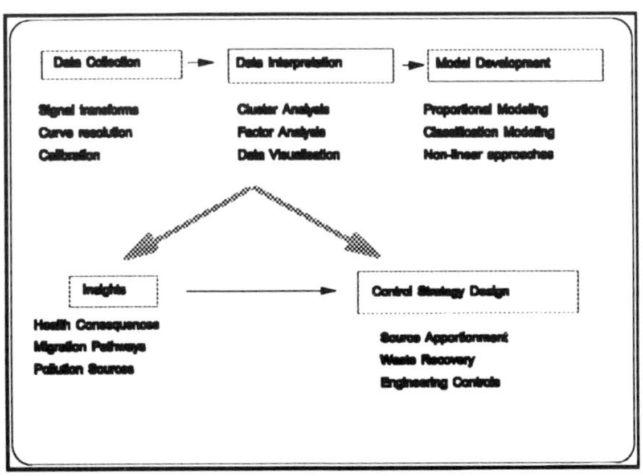

In some instances the environmental chemist may use complex measurement technologies, such as hyphenated techniques (GC-MS, LC-MS, LC-DAD, etc.). These instruments produce two-way data tables per sample, for instance a table of spectra (rows) and chromatograms (columns), which contains information on the concentration levels of several constituents. Advanced chemometric techniques are available for the analysis of such data tables. Specifically when the chromatographic resolution is insufficient, it is possible to retrieve the pure spectra and pure elution profiles by the application of special factor analysis methods, such as Orthogonal Projection Analysis (OPA) [2]. Tables of hyphenated data obtained at several location points and/or at several time slots form a stack of tables, so-called multiway data tables. Advanced chemometrical techniques are available to decompose these tables in pure spectra, pure chromatographic profiles and concentration profiles. This is a relatively new application area, which still has to demonstrate its capabilities to the environmental chemist.

Above methods heavily rely on the principle of linear additivity of signals or linear separability of clusters or categories. Particularly challenging are the non-linear systems which can be visualised and classified by non-linear PCA and QDA. Methods from natural computing techniques, such as Neural Networks have already proven their usefulness in modelling non-linear systems.

A step further is to relate measurements (up to complex multiway data sets) or arrays of results to a certain property. The traditional approach is to apply multivariate calibration (PCR, PLS or MLR) to model these relationships. Recently Neural Networks have been introduced to model complex relationships [3].

In water quality monitoring the operator relies on the quality of the measurements. This calls for intelligent measurement devices which contain internal checking procedures for system suitability tests and internal consistency.

Summarizing, the massiveness and complexity of data sets obtained in water quality monitoring require the application of advanced chemometric and natural computing techniques, preferentially combined in a system of data base management tools, and tools to process heuristic knowledge (if..then rules). Such systems are referred to as data mining for which software is becoming readily available.

2. Current situation

Table 1 summarizes the information which is wanted by the environmental researcher together with the technologies which are generally available to obtain that information. Many of those methods yet do not belong to the toolkit of the end-user and are still in the stage of research and development (techniques indicated with an *)

The wanted information falls into five categories: (1) the analysis of historical (space-time) data (2) methods for improving/cleaning the data (3) techniques to derive concentrations from complex data structures (sensors, hyphenated methods) (4) process monitoring and control (5) thermodynamical/environmental modelling.

Table 1: Data mining techniques relevant to the environmental researcher

Information from data	Data mining techniques
Analysis of historical data	
Visualisation (landscapes)	Principal Components Analysis
Classification (supervised)	K-Nearest Neighbour SIMCA, ALLOC, UNEQ Artificial NN*
Classification (unsupervised)	Clustering, ALLOC Kohonen networks*
Source Allocation/apportionment	Target Transformation FA Artificial NN* PLS
Time/space dependent data arrays	Evolving FA* Three-way FA
Improvement/cleaning of data	
Noise removal	Wavelets* Preprocessing (synchronisation of time arrays*)
Outlier detection	Robust techniques*
Monitoring/control	
Time dependent data arrays (forecasting and modelling) Internal checking of monitoring devices	Multivariate* time series analysis (ARIMA) Multivariate SPC* Kalman filtering* Artificial NN*
Thermodynamic modelling	
Receptor and chemical balance studies	Non-linear modelling Partial Least Squares Simulated Annealing* Artificial NN*
Transform of data to concentration	
Hyphenated techniques	Orthogonal Projection Analysis
Sensors	Curve resolution, three-way FA

The table demonstrates the importance and impact of statistical/chemometrical techniques in the field of environmental chemistry. Because of the specific problems associated with data from eco-systems, highly specialised experts are active in this field, which is known as environmetrics.

Some of the mature chemometric methods are readily available to the end-user, mainly aimed at the visualisation of the data by PCA or by another displaying technique. Software is commercially available in dedicated packages and within general purpose statistical packages such as SAS. However the massiveness of the historical data files may cause problems when using some of the older packages. When exploiting the much larger computing power of workstations, integrated software systems, which are intended to mine the data, so-called data-mining, are becoming available. They offer a choice of modern displaying techniques, factor analysis, artificial neural network, modelling and pattern recognition.

A particular problem with environmental data is the occurrence of outliers, missing data, data below the detection limit etc. This largely influences the ability to extract the information from the data and calls for caution when using 'black box' procedures canned in data mining systems.

Many of the more modern chemometric methods are still in the hands of the specialist and are not common knowledge to the end-user. We refer here to source allocation by target transformation factor analysis and three-way analysis (PARAFAC) of cubes of data. (space*concentration*time). A nice overview of these specialised techniques can be found in a special issue of Chemometrics and Intelligent laboratory systems [9] which reports on the third international conference on environmetrics and chemometrics held in Las Vegas (1995).

In chemometrics literature not much work has been reported on the design of control/monitoring systems, which implies time series analysis, the implementation of forecast and prediction systems, for example by Kalman filtering, and the design of optimal sampling schemes. 'water' quality is inherently an array of quality parameters (constituents and physical parameters), which necessitates the design of systems based on multivariate statistical process control, unless one combines all quality parameters in a single figure, which is the usually approach.
Literature becomes even more scarce when besides objective measurements one wants to include subjective information of the operator in the control system.

3. New developments

The complexity and massiveness of environmental data is a real challenge to the chemometrician and environmetrician. From a chemometrics point of view, there are several areas which need to be further explored. Some of them have been summarized below, without pretending to give a full list :

- Exploration and introduction of data mining technologies to the end-user
- Exploration of three-way techniques and evolving FA for the factor analysis of three way data

- Development and implementation of robust multivariate SPC techniques for water quality monitoring.
- Development and implementation of robust multivariate data analysis (robust for outliers, missing values)
- Development of techniques to analyse non-linear systems and non-linear relationships with non-linear models (Artificial neural nets, etc.)
- Development of techniques for the preprocessing of multivariate data, selection and compression of variables into features, for example by the application of Wavelets.
- Development of measurement devices with internal checks for consistency, automatic fault warning and auto-calibration.

REFERENCES

1. Wenning and G.A. Erickson "Interpretation and analysis of complex environmental data using chemometric methods", Trac 13 446-457, 1994.

2. Cuesta Sanchez, B.G.M. Vandeginste, T.M. Hancewicz and D.L. Massart "Resolution of complex liquid chromatography-Fourier transform Infrared spectroscopic data", Anal. Chem., 69 1477-1484, 1997

3. Hopke, X.-H. Song "The chemical mass balance as a multivariate calibration problem" Chemom. Intell. Lab. Syst. 37 5-14, 1997.

4. Third international conference on environmetrics and chemometrics, Chemom. Intell. Lab. Syst. 37 1-27, 1997.

END USER'S VIEW : STATE OF THE ART, IDENTIFICATION AND PRIORITISATION OF NEEDS.

G. FREGO, ENEL - Residue Valorisation and Treatment Research Center, Struttura Ricerca, Polo Termino, Via Dalmazia 21/c, 72100, Brindisi, Italy

1. Legislation references - the situation in Italy

As far as the monitoring of water quality is concerned, one particular law constitutes the benchmark for the Italian industry - law n° 319 emanated on May 10th, 1976 ("Regulations for the protection of waters from pollution" ; this law is commonly known as the 'Merli law' taken from the name of the parliamentary member who proposed and obtained approval for the bill) .[1]

The Merli law has, amongst others, the following purposes :

- the regulation of any type of waste water, public or private, direct or indirect, in all waters - surface and groundwater, inland and at sea, both public and private, in sewers and on land or in the subsoil;
- the formulation of general criteria for the utilisation and discharge of waters in plants ;
- the organisation of public services such as water supply, sewer systems and depuration,
- the drawing up of a general plan for water remediation on a regional basis ;
- the systematic monitoring of qualitative and quantitative data pertaining to water bodies.

This law establishes a single set of regulations, applicable throughout the nation, for discharges based on the definition of acceptability limits for these as prescribed in tables A, B and C of the law.
Table A, for example, indicates 51 parameters to be respected in terms of concentration limits. This table relates to the limits set for liquid discharges that industries must respect within the first 3 years of the law being approved.
Table C, on the other hand, contains concentration limits for liquid discharges which are more restrictive. Industries are required to conform to these levels within 6 years following approval of the law (this expiry term has been protracted several times).
During the 1980s, therefore, the entire Italian industry had to conform to the limits set for discharges in these two famous tables.
The Merli law set out, moreover, a system of authorisations and monitoring activity to be effected on discharges. "The measuring of discharge shall be performed immediately before the point of immission into the receiving body..."
The law prescribed, amongst other things, a specific controlling authority within the production plants. "The controlling authority is empowered to carry out those inspections,

within the plant, which it considers necessary to ascertain any conditions which give rise to the formation of discharges".
The controlling authority can order that partial discharges containing total toxic metals and non-metals (As - Cd - Cr(VI) - Cu - Hg - Ni - Pb- Se - Zn) undergo special treatment before they run into the general discharge.
With reference to analytical determinations, the Merli law requires that they be effected on an average sample, taken during an interval of not less than three hours.
"The analysis and sampling methods adopted in the determination of parameters are those described in the volume 'Metodi analitici per le acque' (Water analysis methods) published by the The Institute of Water Research of the National Council of Research in Rome (CNR)".
Without going into the details of the complex history of the Merli law and its subsequent modifications and extensions, it is possible to say that this law has been considered as particularly strict, so much so that Italy is exonerated from adopting the European directives on waters.
Only following a sentence issued by the Court of Justice in Luxembourg (dated 28 February 1991) did Italy found itself forced to adopt, from 1992, numerous European directives [previously not considered] under two legislative decrees.
(Legislative decree n° 132 dated 27 January 1992 " Implementation of directive 80/68/EEC concerning the protection of groundwaters from pollution caused by certain hazardous substances" [2]
Legislative decree n° 133 dated 27 January 1992 " Implementation of directives 76/464/EEC, 82/176/EEC, 83/513/EEC, 84/156/EEC, 88/491/EEC, 88/437/EEC, 90/415/EEC on the subject of industrial discharges of hazardous substances in waters ") [3]
These norms gave way to a major review of the requirements established by the Merli law in the matter industrial discharges. From the definition of maximum values for pollutants contained in discharges, applicable throughout the Italian state and for all plants, new specific value limits were set on the basis of each production process (the so-called best available technology) in order to proceed to the next stage which consists of a general system of authorisation that also evaluates the quality of the receiving water body.
Thus, since 1992, Italian legislation on waters has taken a decisive turn towards alignment with European directives. This process of alignment is not fully completed.

2. Design evolution of industrial plants in reference to the norms on protection of groundwaters and industrial plant discharges

The legislative decree n° 132 issued 27 January 1992, which incorporates into Italian legislation the EEC directive n° 80/68 regarding the protection of groundwaters from pollution provoked by certain hazardous substances, imposes particularly stringent constraints on industrial plant design engineers.
In order to obtain authorisation to discharge industrial waste into water bodies it is necessary to provide a documented report which demonstrates that during the design stage everything has been done to ensure that there is no risk of contaminating the groundwaters with substances specified in lists I and II of the attachments contained in directive 80/68/EEC, i.e.

those substances included in the black list and in the grey list of the directive n° 76/464 , which is the source for all European directives on waters.

The system of authorisation, defined in the abovenamed legislative decree, requires that all industrial plants present a geological report, prepared by a qualified geologist, on the hydrogeological conditions of the area concerned, possible depuration capacity of the soil and subsoil, risk of pollution and alteration in the quality of groundwaters.

Similarly, legislative decree n° 133 issued 27 January 1992, which incorporates into Italian legislation 7 EEC directives on the question of industrial discharge of hazardous substances into water bodies, also defines a system of authorisation which involves, with reference to those substances included in the abovementioned black list and grey list, the following :

- the indication of the production capacity of the individual plant;
- the hourly water requirement of the specific production process;
- the possible discharge flow measurement system, where requested;
- the technical equipment employed in the production process and in the waste discharge system to reduce pollution;
- the depuration systems utilised to ensure that emission regulations are observed.

It is important to note, with regard to the Italian situation, that the obligations imposed by the incorporation of the EEC directive, since 1992, have forced industries to carry out extensive analyses and studies in all areas concerning the process fluids used and waste liquids produced. In the large industrial plants the provisions of this law have involved a total reconsideration of the design aspects of all the water cycle.

If one takes, for example, a large ENEL coal-fired power station as a reference the final unified project would take into consideration seven different waste liquid qualities and propose a separate circuit system for each :

- Rainwaters
- Sanitary waters
- Oily water
- Acid and/or Alkaline water
- Waters deriving from coal deposits
- Desulphurisation effluent
- Denitrification effluent

In addition to these seven process waste collection systems it is necessary to consider the principal condenser water cooling system, in which monitoring at discharge of any residual quantity of chlorine compound added for antifouling is often called for .

These references to the current legislation and effects on the design of waste water systems in industrial plants have been made in order to convey the spectrum of problems which Italian industrial
plants have been facing over the last few years , while providing a backdrop for the subject of this paper i.e. the monitoring of water quality.

With these more advanced waste collection system configurations, new forms of waste water management become feasible. It is possible to carry out temporary storage of some wastes so optimising their treatment and reutilization .

The link between various wastes before discharge can be configured in different ways (in series or in parallel). There are technical opportunities (Pinch analysis) which enable an optimisation of the entire plant waste treatment process. Given such an articulated plant system a number of new possibilities arise for the recovery and reutilization of waters (zero discharge).

The problems described above call for an in-plant implementation of waste water measuring instruments capable of monitoring the evolution of the waste water characteristics within the system process (Monitoring of intermediate waste waters).

3. State of the art of water quality monitoring

3. 1. Application fields

In terms of the industrial system, it can be said that the field of application in monitoring of water quality in Italy is defined by the demands imposed the law at the point of discharge (monitoring of output).
In reality the evolution of legislation on the issue, reduction of hydric resources, pollution of groundwaters and increasing supply costs all concur to make industry not only evaluate the quality of discharges but also the quality of incoming water used in the plant.
From this point of view the situation in Italy differs greatly from region to region as a result of its complex geographic relief .
Industrial plants receive their water from groundwater, rivers, lakes and seas. The monitoring of incoming water (monitoring of input) to the plant acquires, in reality, the same importance as monitoring of discharges .
This point leads to the consideration of natural hydric bodies in a new environmental light.
There is an ever increasing number of projects which feature the productive utilisation of the large masses of water contained in the great lakes located in the North of Italy (Lake Maggiore, Lake Como, Lake Garda).
The monitoring of water quality in these lakes constitutes a scientific and environmental issue of great interest which dates back many years with considerable research, especially in the field of water monitoring.
Reference has been made to the importance of monitoring within the plant system (monitoring of intermediate waste waters) .This constitutes, predictably, an application field currently being developed and implemented primarily in relation to water recycling and reutilization policies (zero discharge)
The topic of groundwater monitoring also lies within the sphere of industrial interest because of both the necessity to extract water from existing wells in the plant and to prevent, or control, contamination of the substrata which has taken place during the life of the plant. The dislocation of industrial production residues as a consequence of precipitation has given rise to

the phenomena of leaching and substrata contamination. Industrial companies are often involved in problems of contaminated site remediation . Faced with such events, the health authorities often require the remediation of the site and verification of the results through long term monitoring of the situation (monitoring of contaminated sites).

The same principles apply to sediments that collect in rivers or lakes in the vicinity of industrial discharge outlets and have a long history of pollution phenomena. In these cases monitoring activity also becomes necessary (monitoring of sediment). (see annex 1)
Developing studies and research, monitoring instruments for the large masses of water in the great lakes is also of interest to the Italian industry, which sees in these areas an important hydric resource that is available at certain conditions.

Equally important are the large rivers to which the Italian industry, following law n°183 (1989) under which "hydric basin authorities" were set up, looks with interest in terms of monitoring the water flows, considering these water masses as valuable resources to preserve and prevent from pollution.

3. 2. Measurement aim

The measuring aims in the activities indicated above are, from an industrial point of view, first and foremost observance of the law and compliance with requests from controlling authorities (monitoring of output).
The economic necessity to optimise the management of wastes forms a second aim (monitoring of input, monitoring of intermediate waste water).

The need for environmental control, often imposed on industry by health authorities, constitutes a commitment which requires the development of monitoring activities that accompany remediation activities in cases where the environmental situation is endangered (monitoring of groundwater, monitoring of contaminated sites, monitoring of sediments)

3. 3. Measured parameters and considered concepts

The industrial measurement needs illustrated above necessitate the development of diverse instrumentation.

3. 3. 1. Monitoring of output

As far as the first need is concerned i.e. waste control (monitoring of output), the ranges to be measured are substantially those comprised within the European Directive lists (black and grey lists contained in directive 76/464 and attachment I of Directive 80/778).
The legal control of wastes requires considerable reliability from the instrumentation utilised in continuous monitoring .
The discharge of cooling water from industrial plants such as Thermoeletric Power Stations can create ecological problems.

For example flow of cooling water (80 m^3/sec) from the Porto Tolle (4x660 MW) Thermoelectric Power Plant can be taken and discharged in River Po, the Adriatic sea or even the nearby Lagoons (see Annex 2).
Monitoring of water and ecological conditions must be studied with care and assessed for environmental impact.
The study of ecological impact calls for the development of many research lines (see Annex 3).
The quality of water is only one the the many aspects to be considered.
Monitoring activity can be extended to an entire river tract. [4] [5]
That is the case for the River Po (see Annex 4) where eight Thermoeletric Power Plants, located along the River, make it necessary to create a network of continuous monitoring of water quality (dissolved O_2, Ph, conductivity, turbidity, temperature) (see Annex 4).

3. 3. 2. Monitoring of intermediate waste waters

The requirements of different industries vary greatly . Briefly, it can be said that it is fundamental to develop instrumentation able to monitor the result of the principal depuration treatment processes utilised in industry.
These include:

- sedimentation processes
- clarification processes
- precipitation processes
- defiltration processes
- disinfecting processes
- biological deposit elimination processes

3. 3. 3. Monitoring of groundwater

In some situations it becomes necessary to define programs for the monitoring of groundwaters subject to risk of contamination.
It is important to develop measurement techniques capable of providing reliable instruments and measurements which are easily installed for long term measuring campaigns . Further development of sensors, other than those for traditional parameters (temperature, pH, conductivity, dissolved oxygen, Redox potential, turbidity), should include those relative to the list of substances contained in the attachments of the abovementioned European Directives on water quality. The configurations of measurement and performance must be adapted to the monitoring procedures referred to autonomous units with reduced maintenance needs, long operation times and continuous data recording.
Groundwater monitoring and characterisation activities have resulted in the construction of penetrometric units capable of collecting soil, interstitial air and water samples, at the same time. (see Annex 6).
With reference to water, it is possible to measure Ph, Redox potential, conductivity, dissolved O_2, temperature and pressure).

Many other parameters could be measured in this manner utilising the miniaturisation of the measurement flow cell.
Mobile laboratories can utilize this penetrometric unit to characterise contaminated sites "in situ".

3. 3. 4. Monitoring of sediment

Previous experience leads us to propose and sustain the development of direct measuring techniques capable of supplying information on the condition of the sediment present on sea and lake beds which have been subjected to pollution.
Measurements concerning the water-sediment interface must have a high spatial elevation resolution (millimetric) given the need to calculate strong gradients . They must also cover a sediment depth which is representative and variable according to the parameter measured.
The water-sediment interface is a key site for all the transformation processes that take place within the sediment ; it also acts as the exchange point between the sediment system, interstitial waters and/or depth waters (nitrification/denitrification, sulphate reduction and sulphur precipitation, phosphate and heavy metal adsorption/desorption, molecular diffusion flows from and to sediment etc.).
The following parameters, amongst others, are indicated for measurement :

- dissolved oxygen
- pH
- Redox potential
- sulphurs
- conductivity
- ammonia
- nitrates
- hydrocarbons

Each of the parameters listed above, along with the relative sensors, presents a different measuring and operating complexity.
Whatever the case, all concern sensors of limited dimensions (from about 10 to 5000 micrometers) which are placed in the sediment at speeds which vary according to the sensor response time and at depths beyond which any information acquired is considered to be of secondary importance. (see annex 7)

3. 3. 5. Monitoring of contaminated sites

The issue of monitoring of contaminated sites is one of great interest to industry.
Work in this area includes the characterisation of sites using mobile laboratories that can supply rapid responses to questions of risk assessment connected to the phenomenon of water, soil and air pollution.

In terms of water, account must be taken of this aspect in the development of rapid leachate analyses. The parameters to be considered here are also those listed in the attachments to the earlier mentioned European Directives .

In this case the measuring techniques must provide responses in real time in order to achieve the aim of the measurement campaign and the immediate analyses of the samples taken in the site under evaluation. [4]

It is interesting to develop biosensors, capable of producing fast and low cost "in situ" measurements, for the monitorinf of pollution in contaminated water.

An immunochemical biosensor for the determination of PCB concentration in contaminated water is currently under development within the terms of a project defined by the Environment Programme founded by the European Union. (contract n° eV5V CT - 94 - 0407 Monitoring of Polychlorinated byphenyls (PCB) in Groundwater through an immunochemical Bionsensor with electrochemical detection).

System characteristic include :

- ease and rapidy of performing contaminated soil and water analyses;
- possibility of performing determination "in situ" directly in line with the and without the necessity to treat the samples (see annex 8).

3. 4. Present performance

It is very difficult to express an opinion on the Italian situation in this field.

There is much perplexity on the quality of the instrumentation utilised in continuous monitoring.

The acquisition of measurements, biological growth on instruments, the frequent need to calibrate and reset sensors lead us to conclude that there is still much work to be done in order to implement the quality of information that continuous monitoring systems are able to provide.

3. 5. Main drawbacks and limitations

- Reliability over time of the continuous measurement of water control parameters is a critical factor in many industrial plant situations
- The control methodologies are referred to laboratory analysis methods which involve tests on samples taken from waste waters in a non-continuous manner .
- For reasons of uniformity, existing legislation often only indicates a single method of measurement (measurement of samples taken from the plant and analysed in the laboratory).

These last two points do not encourage the development and acceptance of instrumentation which functions on a continuous basis and inhibits the development of instruments using innovative methodology.

4. Identification and prioritisation of technical and regulatory needs

Experience has underlined the need for technical norms in the area of industrial wastes which favour the utilisation of continuous instrumentation; conversely the norms tend to restrict the development of new instruments .
A paramount example of the former is the American regulation issued by the ASTM (American Society for Testing and Materials) which, for many of the parameters relative to precisely water and environmental technology, include methods that are specific and exemplified for on-line monitoring .

4. 1. Classical parameters : required performance

With reference to the classical parameters reported in the attachments to the European Directives on waters, it is proposed that a set of technical regulations be drawn up which promotes on-line monitoring and recognises, furthermore, the possibility of developing new measuring techniques .
On this point the following are considered important in terms of interesting techniques :

- FIA technique (Flow Injection Analysis)
- PSA technique (Potentiometric Stripping Anodic)
- LFA technique (Loop Flow Analysis)

4. 2. New concepts to consider , new parameters to be measured

The monitoring of groundwaters, contaminated sites and sediments requires the development of new measurement parameters and new instrumentation.
The development of mobile laboratories for surveys of contaminated sites emphasises the importance of creating instrumentation which is able to offer rapid screening of the contaminated situation and facilitate risk assessment of the area in question.
In this case an evaluation of the need to develop biosensors and, more specifically, techniques which can give a global evaluation of toxicity, must be made.

5. Conclusions

In terms of the situation in Italy, industry's attention to monitoring of water quality is becoming increasingly conditioned by legislation and, more particularly, by the recent integration of European norms on water quality

The authorisation procedures, prescribed by European norms and directives, require declaration of production cycles, their capacity and concentrations. Supervision by health authorities is not only confined to the output discharges from industrial plants but also within the same plant and may even extend to special prescriptions for partial internal discharges.

In relation to the various forms of hydric supply and specifically those being currently developed (supply from hydric basins, rivers, great lakes) it is in the interests of industry to develop continuous water quality monitoring systems for use at the plant inlet. This will enable plants to modify, control and optimise the initial treatment costs of incoming waters and control the phenomenon of sudden pollution of the water supply source .

With reference to past experience acquired in the field of possible applications and implementations for water quality monitoring it is important to extend these to the monitoring of sediments contained in the hydric bodies receiving the industrial discharges and to the monitoring of leachates (leaching of rain waters on the residues) which may be associated to industrial activity.

Industrial production involves the continual flow of wastes and the need for automatic controls and regulation of treatment plants; it is therefore essential to have adequate monitoring instrumentation. The reliability of such instrumentation is an issue of great concern in the industrial field. All new research initiatives are welcomed, as are the funds necessary to develop monitoring measurements and techniques which are highly reliable and applicable to the control and regulation of industrial treatment systems for polluted waters.

The analytical methods indicated in existing legislation and directives must also take into consideration the need for rapid response and privilege techniques and methods capable of providing immediate results.

Examples of future promising techniques are indicated below :

- FIA technique (Flow Injection Analysis)
- PSA technique (Potentiometric Stripping Anodic)
- LFA technique (Loop Flow Analysis)

In terms of new approaches and ideas to be developed in the field of biosensors the following are pertinent to :
- global evaluation of toxicity

The speed of response provided by biosensors deserves careful attention given the foreseeable advantages offered providing that the measurement systems are thoroughly validated and reliable.

In the context of the Standard, Measurements and Testing, financially supported research aimed at obtaining instrumentation capable of supplying reliable results in minimum time should be considered as a route to followed with greater commitment and resources so as to confer a competitive advantage to European industry and improve environmental protection.

6. List of annexes

- Annex 1 : Monitoring of water quality application fields
- Annex 2 : Porto tolle - thermoelectric power plant (4x660 mw)
  Location of the plant and sketch of the cooling circuit
- Annex 3 : The research lines of the environmental impact assessment
- Annex 4 : ENEL/CRAM - network for the continuos monitoring of poriver water quality
- Annex 5 : Monitoring of output "aqua" automatic unit sensor characteristcs
- Annex 6 : Monitoring of groundwater illustration of sampling lines and analyses
- Annex 7 : Monitoring of sediment
- Annex 8 : Monitoring of wastewater

REFERENCES

[1] "Regulations for the protection of waters from pollution", Legge n° 319 - 10 maggio 1976

[2] "Implementation of Directive 80/68/EEC concerning the protection of groundwaters from pollution caused by certain hazardous substances", Decreto Legislativo n° 132 del 27 gennaio 1992

[3] Implementation of directives 76/464/EEC, 82/176/EEC, 83/513/EEC, 84/156/EEC, 88/491/EEC, 88/437/EEC, 90/415/EEC on the subject of industrial discharges of hazardous substances in waters", Decreto Legislativo n° 133 del 27 gennaio 1992

[4] Environmental impact assessment for Italian thermoeletric power plants in coastal zones by R. Ambrogi (ENEL -DSR) and R. Vitali (ENEL -DCO), Bullettin de l'Istitute Oceanographique (Monaco - 1992)

[5] L'influenza del Lambro sul Po., Analisi dei dati della rete di monitoraggio in continuo dell' ENEL, L. Guzzi, G. Bonforte, A. Bozzani, W. Martinotti, ENEL SpA - Centro Ricerca Ambiente e Materiali Milano (IRSA 1996)

[6] Characterisation and monitoring of industrially contaminated land: The ENEL approach by Giuseppe Frego, Pompilio Caramuscio, Riccardo Elini, Giuseppe Varallo - ENEL SpA-DSR Residue Valorisation and Treatment research Centre – Brindisi, (Network Industrally Contaminated Land in Europe) - Inaugural Meeting of NICOLE, Hannover 21 May 1996.

Annex 1 : Monitoring of water quality application fields

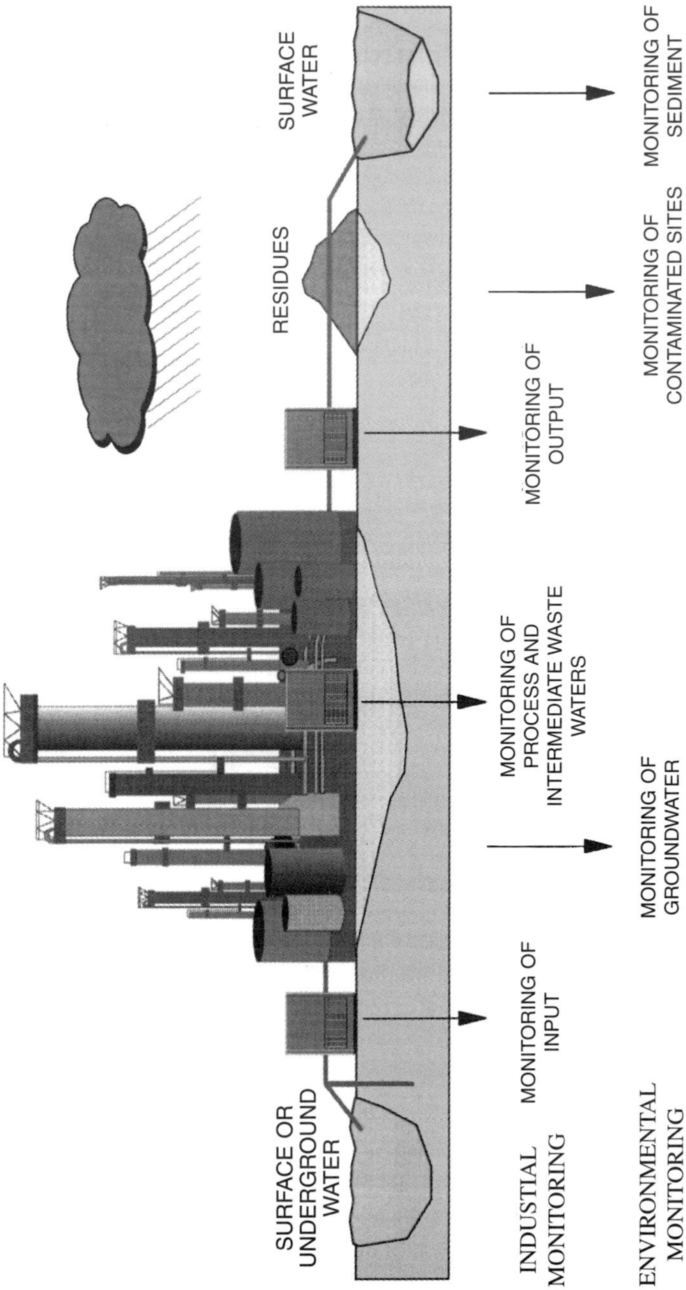

Annex 2 : Porto tolle - thermoelectric power plant (4x660 mw). Location of the plant and sketch of the cooling circuit

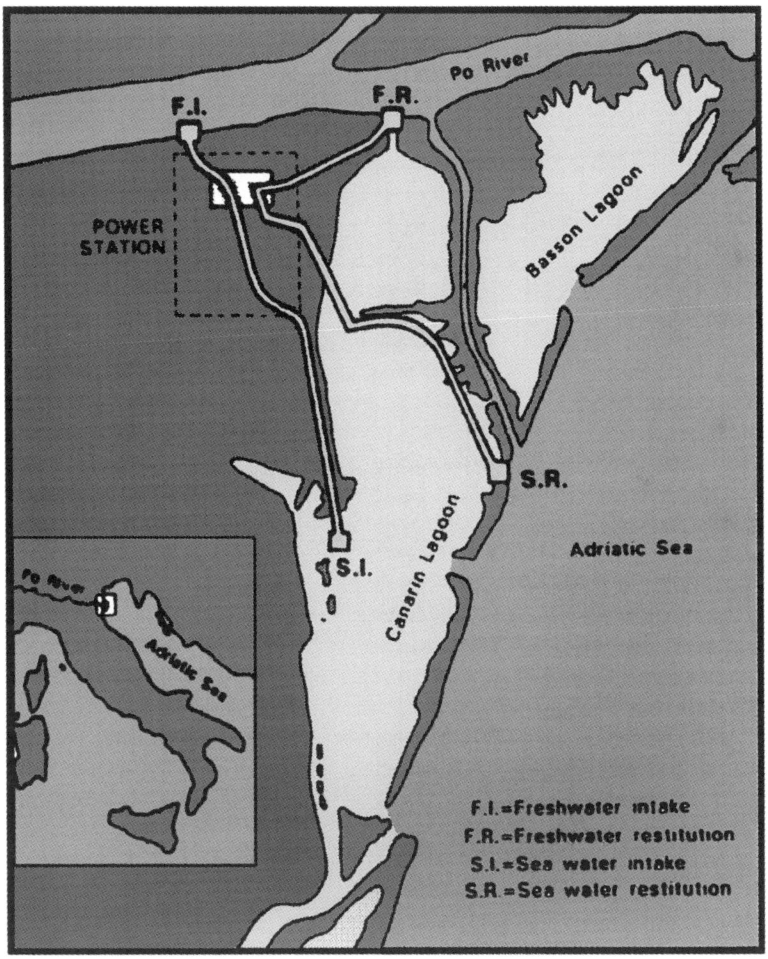

Annex 3 : The research lines of the environmental impact assessment

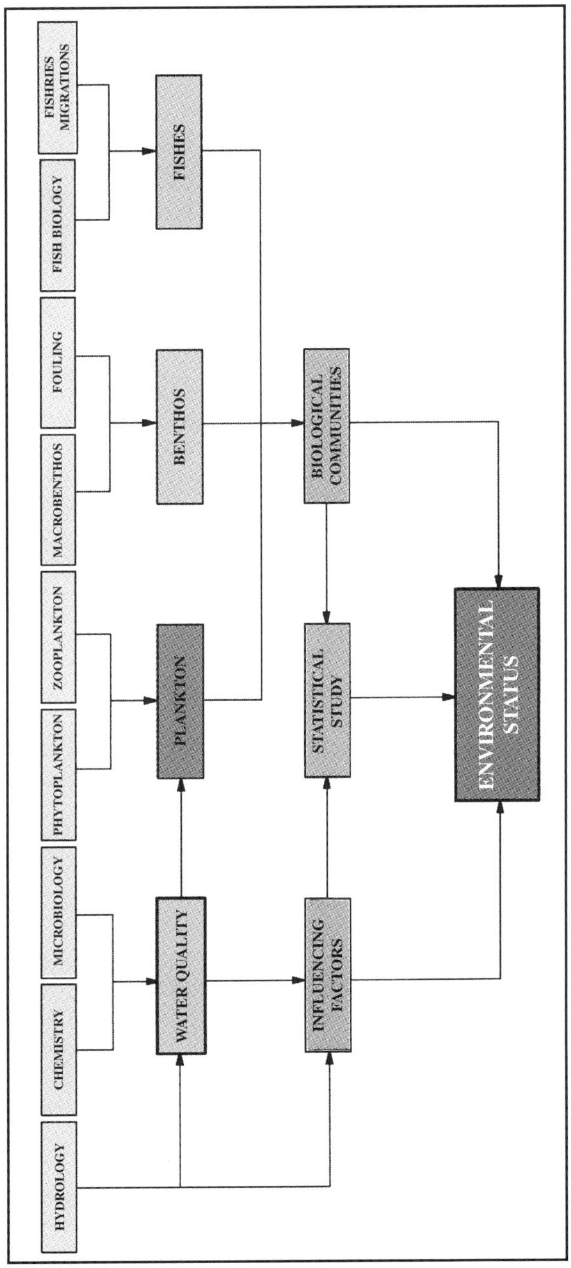

Annex 4 : ENEL/CRAM - network for the continuos monitoring of poriver water quality

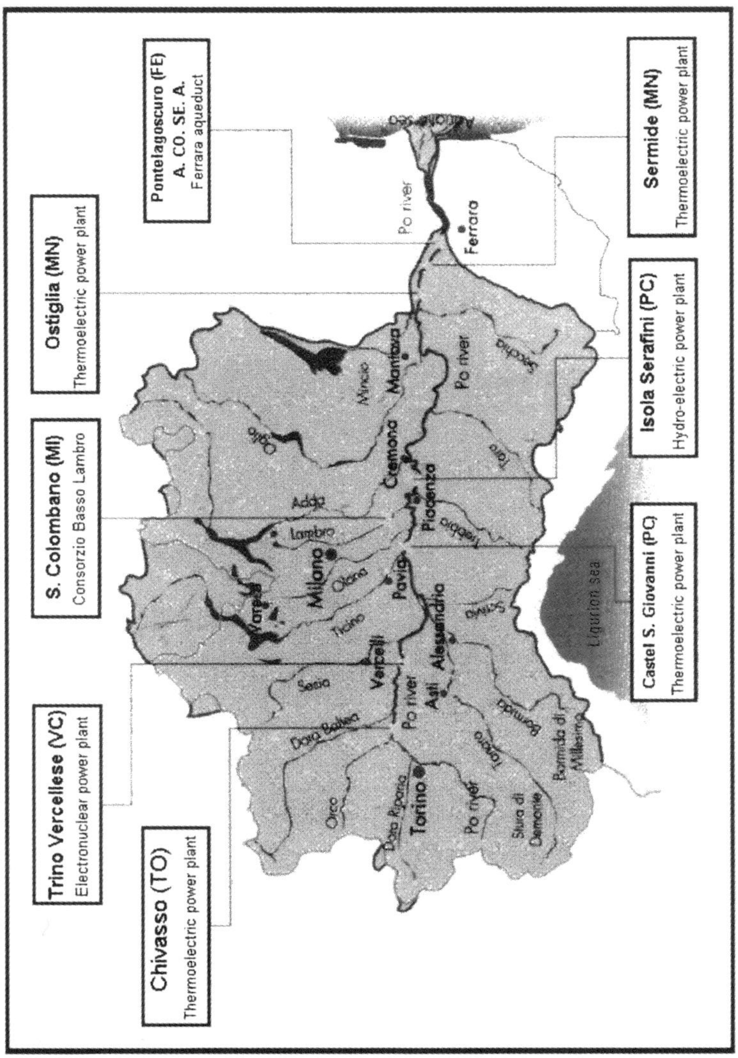

Annex 5 : Monitoring of output "aqua" automatic unit sensor characteristcs

PARAMETER	RANGE
temperature (°C)	-5 ÷ +35
conductivity (ms/cm)	0 ÷ 6
pH	0 ÷ 14
redox (mv)	± 1250
oxygen (mg/l)	0 ÷ 21
turbidity (FTU)	0 ÷ 2/20/200/2000
hydrocarbon fluorescence	0 ÷ 9999.9
well temperature (°C)	0 ÷ +50
piezometric level	0 ÷ 2 (bar)

Annex 6 : Monitoring of groundwater illustration of sampling lines and analyses

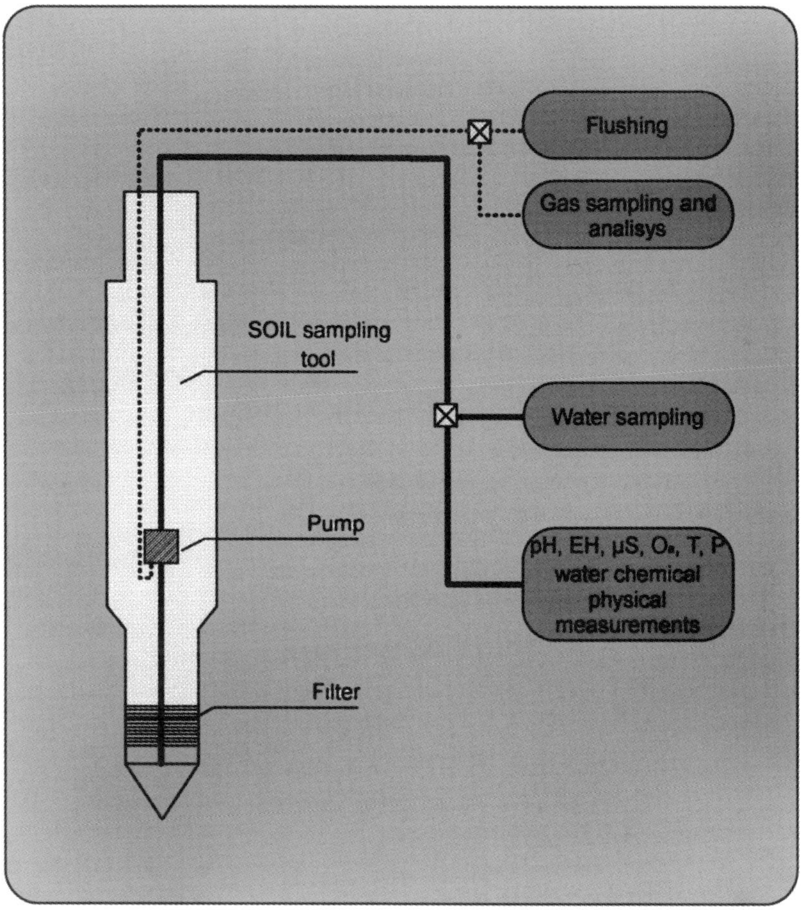

Annex 7 : Monitoring of sediment

SWIMP - sediment water interface microprofiler

Sulphide microsensor

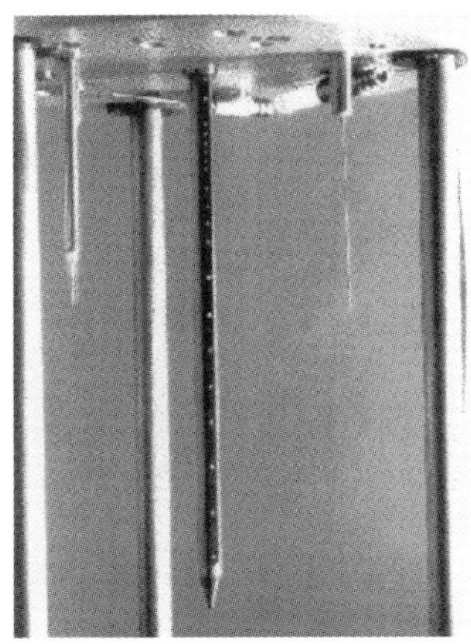

Annex 8 : Monitoring of wastewater

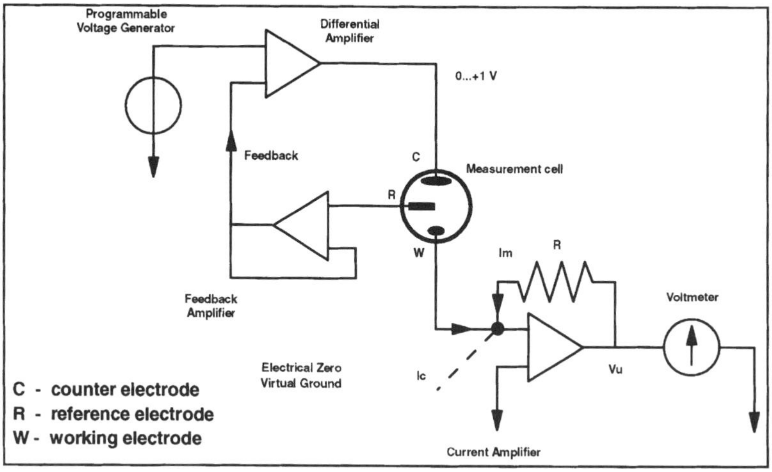

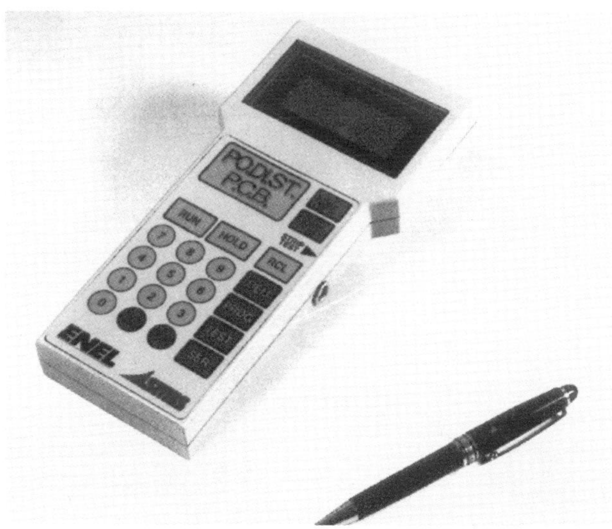

CHARACTERIZATION OF WASTEWATER SLUDGES END USER'S VIEW

A. KOUZELI - KATSIRI, National Technical University of Athens, Iroon Polytechniou, GR 15773 Zographou, Athens, Greece.

1. Summary

This paper gives an overview of the state of the art on the problem of sludge characterisation. The methods considered are related to sludge production, handling, thickening, dewatering, stabilisation, incineration, agricultural use and landfilling. Measured parameters and concepts are considered and the main limitations and drawbacks are discussed.

2. Introduction

Sludges from different sewage works can exhibit wide differences in their physical, chemical and biological characteristics, and there can also be seasonal variations in the sludge characteristics at a particular works. In practice, this means that the designer of a sludge treatment plant should initially obtain information about some specific properties of the sludge to be treated, and about any temporal variations that might occur in sludge properties, in order to produce the best overall process design specification.

It would certainly be unwise to assume that all sludges respond to a given treatment process in the same way. For, example, some sludges may not digest satisfactorily, while others may not dewater adequately even with the use of coagulants, so that the "blind" application of digestion or mechanical dewatering in such cases could lead to expensive process failures.

Characterisation tests to determine "digestibility" and "dewaterability" would prevent such a problem. Similarly, the plant operator often needs to undertake daily laboratory characterisation tests to enable him to identify the optimum operating conditions for the processes treating the plant sludge.

Similar remarks can be made as regards sludge disposal methods. If sludge will be disposed to tip for example, it is necessary to know several mechanical properties of sludge such as its shear strength, its compressibility etc. tests are therefore necessary and the availability of laboratory or pilot equipment to conduct them is equally important. Table 1 gives an overview of the main sludge characteristics related to processing and handling of sewage sludges.

Table 1 : Characterization tests in relation to sludge treatment and disposal methods

Method of sludge treatment and disposal	Transportation	Sedimentation	Activated sludge	Stabilization aerobic	Anaerobic digestion	Chemical stabilazion	Thermal	Thickening by gravity	Flotation	Mechanical methods	Drying beds	Mechanical methods	Tipping, barging	Composting	Incineration	Agricultural use
Temperature				+	+					(+)	+	+				
Density	+															
Dynamic viscosity	+															
Rheological properties	+										+	+				
Settling behaviour	(+)	+		+	+	+	+	+								
Sludge volume/sludge index		+	+					+								
Moisture content/total solids residue		+	+	+	+	+	+	+	+	+	+	+	+	+	+	(+)
Volatile substance/fixed residue			(+)	+	+	+	+				+	+	+	+	+	+
Organic carbon			(+)	(+)	(+)	(+)	(+)						(+)	+	+	(+)
COD of liquid phase					+											
Digestibility	+				+											
Digestion behaviour					+											
Digester gas composition					+											
pH value	+	+		+	+	+					+		+		+	+
Acid consumption					+											
Volatile acids					+											
Fats and oils				+	+								+			(+)
Surfactants					+				(+)				+			(+)
Heavy metals					+								+	+		+
Other pollutants (cyanides, halogenated hydrocarbons etc.)		+		+	+								+	(+)		+
Nutrients (N, P)		+		+												
Enzymatic activity		(+)		(+)												
Conditionability					+	+						+				
Particle size distribution				+	+		+					+				
Dewaterability				+	+	+	+					+				
Capillary suction time (CST)								+	+	+		+				
Specific resistance to filtration								+	+	+		+				
Compressibility												+				
Centrifugability												+				
Calorific value				(+)	(+)		(+)								+	
Heating by water														+		
Leachability														+		
Microbiological							+							+		+

In the framework of COST 681, WP1 was particularly concerned with the development of improved tests (1) to determine key physical and chemical characteristics of sewage sludge. Altogether 28 different parameters were identified and published in a compendium of methods (6,7).

These parameters are listed in table 2. Less than half of these methods are standard methods, about one third are tentative methods only and 12 methods are still at the exploratory stage. Thus there is some way to go before a full suite of standardised and harmonised methods of testing sludges is available to the design engineer. In the following discussion parameters will be grouped according to their relation to the sludge treatment processes or disposal method.

3. Sludge production

Within a particular sewage treatment works sludge production will be influenced by the volume of incoming wastewater and its quality. The latter will be influenced by the type of activities and the district served. Within the treatment plant sludge is produced by primary sedimentation, by biological treatment processes and by chemical processes, (2), (5).

Sludge production by biological treatment processes will depend on the organic load of the plant but also on the type of the process used. For example extended aeration processes give very low sludge production rates, due to auto-oxidation by bacteria. Similarly chemical precipitates depend on the dose and type of the coagulant added as well as the volume of the influent. Therefore it is very difficult to arrive at sludge production values in terms of per capita or g per DS of incoming sewage. There is therefore a need for these parameters to be accurately determined in each plant.

A protocol for the measurement of sludge production at sewage treatment plants has been proposed by Bruce (4). Both mass and volume of sludge must be measured. On line flow meters using mostly magnetic methods have been used successfully for measuring sludge volumes. Solid concentration can also be measured on-line using optical analysers or ultrasonic attenuation techniques.

Table 2 List of parameters included in the Compendium of Methods for the of sewage sludges in relation to the design and operation of sludge treatment processes

Gravimetric	**Stabilisation**
1. Total solids (percent dry matter)	17. Anaerobic digestibility
2. Volatile solids (loss on ignition)	18. Aerobic digestibility
3. Dissolved solids	19. Volatile fatty acids
4. Suspended solids	20. Bicarbonate alkalinity
	21. Digester gas composition
Consolidation	**Stability**
5. Column tests and centrifuge tests	22. Stability index
	23. Odour intensity
Conditioning and dewatering	
6. Conditionability (coagulant demand)	**Incineration**
7. Effect of shear	
8. Specific resistance to filtration	24. Calorific value
9. Suction time (Buchner funnel)	
10.MFT test	**Handling and disposal**
11.Capillary suction time (CST)	
12.Compressibility coefficient	25. Rheological properties
13.Ultimate cake solids content	26. Physical state
14.Particle mobility	
15.Filter leaf yield	**Other parameters**
16.Centrifugability	
	27.Particle size distribution
	28. Density

4. Thickening

Thickening is a process aiming at reducing sludge volumes up to 4 times. Both undigested and digested sludges are subjected to thickening but probably, the most cost effective single process in sludge treatment is gravity thickening of raw sludge since volume reduction has such a profound effect on downstream processing costs, and obviously on transport. A number of tests have been used to estimate thickening performance.
These are listed in table 3.

SVI and SSVI have found widespread use particularly in the control of the activated sludge process. These methods are based on column batch experimental results. The diameter of the column must be large enough to prevent wall effects and other artefacts. Low speed stirring used in the second method limits these problems.
Dimensioning of thickeners is often made by using the flux method which calculates the maximum possible flux of suspended matter through the most critical section of the thickener (26) (38).

A more sophisticated design procedure for sludge consolidation tests was described by Hoyland and Day (11). For any particular sludge, tests are performed using a small centrifuge and a pilot-scale thickener to evaluate three basic parameters (compression coefficient, compression index and resistance to consolidation) which characterises the sludge consolidation properties. Predictions made by the model indicate that, depending on the size of the tank and the resistance to consolidation, there is an optimum height of tank which maximises the concentration of solids in the consolidated sludge.
This research has lead to a dramatic increase in the use of consolidation in the UK and to a lesser extent elsewhere. Similar tests using a low speed centrifuge have been described by Colin and co-workers (36, 37).

Table 3 : Thickening tests

Parameter	Measurement
Sludge Volume Index (SVI)	Sludge volume/suspended solids (ml/g)
Stirred Sludge Volume Index (SSVI)	SVI in a slowly stirred graduated cylinder (ml/g)
Low speed centrifugation	Sludge blanket settling velocity in a field of 10-100 G
Ultimate cake solids content	Dry solids concentration after settling or centrifugation.

5. Dewatering

Sludge dewatering is one of the main unit processes in sludge treatment as it minimises sludge volumes for final disposal. In some countries i.e. Germany, authorities require a high solids percentage, over 35%, if sludge is to be disposed of by controlled tipping. Sludge dewatering can be achieved by filtration or centrifugation. Prior to dewatering sludge is always conditioned.

The parameters related to conditioning are measurements of dewaterability in relation to the amount of coagulants. The most important tests related to conditioning and dewatering are listed in table 4.

Filter leaf testing relates directly to vacuum filters and is quite uncommon in Europe apart from France.

The first two parameters r and CST have found widespread use and acceptance and are by far the most commonly quoted indicators of dewaterability, both measured under standardised conditions (1), (20), (39), (21), (22).

An innovative procedure for automatic measurement of specific resistance has recently been reported by Springer (19). Complementary to specific resistance to filtration are the compressibility coefficient and the shear strength tests.

Methods for predicting belt press performance have been proposed by several authors. Heide and Kampfe (8) described a new procedure for characterisation of the dewatering behaviour of sludges in belt presses, for selecting suitable polyelectrolyte coagulants and for optimising the use of selected polyelectrolytes. The procedure evolved from laboratory scale and full scale evaluations and comprises a combination of conditioning at laboratory scale and the application of a single dewatering test - the modified filtration test (MFT) - which gives an indication of dewaterability as well as the attainable dry solids content of the cake. Verification of the procedure was demonstrated at more than 10 sewage works where aerobically digested sludges were dewatered by belt presses.

Another common method of dewatering is centrifugation but laboratory assessment of sludge "centrifugability" has long proved difficult. Spinosa and Mininni (9) concluded that centrifugability characterisation is very difficult as it is practically impossible to reproduce in the laboratory the conditions actually existing in an industrial centrifuge. In industrial centrifugation three properties are important: settleability, scrollability and floc strength.

Unfortunately no standard methods are available in which the above methods, are considered as a whole. Two methods can be used to evaluate these properties and to select the best type and dosage of polyelectrolyte. In the first method a bench centrifuge is used and the SS concentration is measured after centrifugation, while scrollability is determined by a penetrometer test on the cake collected in the tube. Optimal polymer dosage can be determined and cake concentration and solids recovery can be predicted. However, the procedure is difficult to apply to activated sludges because of their poor consistency.

The other, more promising, method is based on floc strength measurements (based on the COST 68 standard procedure (1) and these seem to provide a reliable means of determining optimal conditioner dose.

Despite the availability of these tests many designers would prefer to carry out full scale trials to evaluate centrifuge performance.

One of the successful new developments in methods for assessing sludge "conditionability" and optimum conditioner dose was described by Cambell and Crescuolo (10), (45) and it has lead to the design of an automatic on-line control system for the dosing of polyelectrolytes.
The basis of the method is the continuous measurement of the rheological properties of sludge using a co-axial viscometer interfaced with a computer to analyse the rheogram and ultimately to proceed a feed-back control loop. Using a standard test methodology of this type, various sludges tested showed a distinct peak in the rheogram at a particular polymer dose and this enabled the optimum dosage to be identified.

6. Stabilisation

The object of all stabilization processes are to make sludge less offensive. The first definition of aerobically stabilised sludge proposed by COST - 68 (1) was a sludge where the odour intensity Index (OII) does not exceed 11 at any time prior to 14 days of storage at 20 °C unless the odour can be classified as a typical soil odour. However odour is a highly individual perception and its measurement requires a panel of test persons and a rather laborious test procedure. Odour measurements have therefore found little practical application in sludge treatment.

The ability to assess the anaerobic digestibility of sludges in the laboratory is important for design engineers and plant personnel since such information is useful in predicting gas production and volatile solids reduction. It is also very important to carry out laboratory tests before embarking on the design of digestion facilities because sludges vary in their amenability to digestion. Occasionally, sludge from a particular works will not digest satisfactorily and if initial tests had not been made before a full-scale plant was installed this could represent an abortive capital investment.

Table 4 Dewatering tests

Parameter	Measurement
Specific resistance to filtration (r)	Rate of solids buildup in a filter cake
Capillary suction time (CST)	Rate of liquid withdrawal from a sludge by capillary suction

Parameter	**Measurement**
Compressibility	Rate of increase in r with increasing pressure in filtration
Filter leaf yield	Sludge solids captured per unit area in a vacuum filter
Shear strength	Rate of change in r or CST with turbulent strength on sludge particles
Filter belt press yield	• Rate of gravity water drainage from sludge on filter bed • Rate and extend of water removal from sludge subjected to compression subsequent to drainage

There are various laboratory digestion tests described in the literature and itemised in the COST Compendium Methods (6). Stuckey (15) introduced the 'Biochemical methane potentials' (BMP) test which involves incubation of test and control samples in 250 ml reagent bottles stoppered with rubber serum caps. Gas production during incubation is monitored using a wet syringe and the samples of gas can also be analysed for methane content. Such tests allow routine monitoring of sludge biodegradability and also reveal the presence of any components toxic to methanogenesis.

The method proposed by the UK Department of the Environment (40) is based on the comparison of gas production of the sample sludge and that of a reference sludge (generally a raw non-toxic sludge with total solids content higher than 4%). The seed sludge is obtained by a laboratory reactor fed with reference sludge (detention time 24 d). The test is carried out with bottles containing different ratios of reference and sample sludge to seed sludge (1:10, 1:3 and 1:2 respectively) at 32°C. The gas produced is recorded twice a day for five days. Three amenability factors are then calculated which represent the ratios of the actual gas yield of the sample sludge to that of reference sludge at the same dilution with seed sludge. The presence of toxics, and possibly a typical behaviour can be so evidenced.

A rapid test lasting 22 hours is also proposed by the UK Department of the Environment.
Experiences of sludge digestibility were also carried out at the Italian Water Research Institute. The proposed method is quite similar to the previous one, but the reference sludge is not considered due to the difficulties of providing and/or producing such a sludge in controlled conditions. The test is carried out at 37°C with bottles filled with sample sludge and seed sludge at different ratios (from 1:1 to 1:5). The gas yield is recorded once a day and the test lasts until a marked decrease is observed, in any case no more than 30 days.

In the Federal Republic of Germany a DIN standard named "Amenability to anaerobic digestion" is existing since 1985 (46). It is based on measurements of gas production of mixtures of the sample sludge with an almost fully digested sludge as seed sludge compared with mixtures of a reference sludge (raw sludge of municipal origin) with the seed sludge. The test is carried out at 35°C over a test period of at least 20 days. The results are expressed as sum of particular specific gas production in litres per kg organic mass (determined by loss on ignition) against the associated test duration in days.

A method of toxicity evaluation using micro-algae has recently been demonstrated by Pun and co-workers (27)

The biochemical "stability" of a sludge is an important parameter but so far there is no standard method for quantifying "stability". An unstable sludge can change in its character very rapidly during anaerobic storage and, for example, show a dramatic increase in coagulant demand. It will usually also become more offensive in odour. Microbial activity measurement for anaerobic sludge digestion has been reported by Chung (33).

By contrast, a completely stable sludge, by definition, will show no adverse changes on storage. Bruce (16) described a variety of non-standard tests to assess the "stability" of both raw and digested sludges (odour potential, volatile solids fraction, biodegradable residue, chemical composition, biological activity).

He also described a test for assessing the stability of raw sludges developed at the Water Research Center (now WRc). This test involves anaerobic storage of the sludge in a standard stirred cylinder at a constant temperature (usually 208C). The sludge is stirred slowly and samples are withdrawn periodically for coagulant demand, non-settleable solids (fines) and odour intensity.

Oxygen uptake rate has been proposed as an indicator of sludge stability in aerobic processes (excess activated sludge, aerobic digestion). In this field great progress has been made and several on-line methods for rapid sludge characterisation have been reported (28), (42), (43), (47).

Moreover several tests exist which measure various types of enzymatic activity in activated sludges. A low activity level indicates a well stabilised sludge (30), (24), (25), (23), (31).

The major application of on-line oxygen uptake measurement is process control as it allows the optimisation of the activated sludge process and the control of the influence of toxic effects. The data of two oxygen electrodes are processed continuously using a microprocessor.

7. Physical and rheological properties

The physical properties of sewage sludge are especially important for the design of sludge pumps, thickeners and dewatering equipment, for the optimisation of conditioning, dewatering, heat treatment and landfilling operations.

One of the earliest major contributions to the COST 68/681 activity was the work by Colin (12) on methods for characterising the physical state of sludges. A knowledge of just the water content of a sludge is quite inadequate to predict precisely its physical state, which may be liquid, plastic, solid state with shrinkage or solid state without shrinkage. The contribution made by Colin was to introduce a suite of laboratory tests which allow the physical state of a sludge to be defined quantitatively.

Tests are made using viscometers, penetrometers, extrusion equipment and deformation measuring techniques. Physical properties of sludge and relative tests are shown in table 5.

Colin later described some extensive studies using lysimeters and pilot columns to observe the changes which occur in sludges in simulated conditions of a landfill, both the physical and chemical changes in sludge characteristics being observed, and the effect these changes might have on both the landfill structure and any leachate emanating from it. Guezens and Dieltjens (14) also described some methods for determining the mechanical strength of cohesive sludges in relation to their deposition in landfills.

Another fundamental contribution in the area of sludge characterisation was by Frost (18) who showed that there is an inter-relationship between the rheological behaviour of a sludge and its solids-liquid separation behaviour (i.e. settleability, thickenability and dewaterability). The relationship between viscosity and nominal compressive pressure NCP is much stronger than that between viscosity and solids concentration, by a factor of 25.

In essence, Frost propounded that the application of a given NPC to different sludges would produce sludges with similar viscosity but very different solids concentrations. Frost concluded that the explanation for the inter-relationships may lie in the bound water content of sludges and that under a given compressive pressure sludges will thicken to about the same volumetric concentration of particles.
The practical application of this finding is the ability to produce a sludge consistency scale (analogous to a pH scale) which could be used to identify the optimum sludge concentrations for specific treatment and disposal routes.
This latter work is ripe for further study, with the prospect of identifying a key parameter from which all sludges could be fundamentally characterised in relation to physical behaviour.

8. Thermal properties

Increasing attention is paid nowadays towards final treatments of sewage sludges like drying, incineration, pyrolysis. The main characteristic connected with these processes is the calorific value of solid and liquid fuels. The experimental technique consists of burning a sample in the bomb filled with pure oxygen and measuring the amount of energy transferred to the calorimeter.

This is possible by determining the temperature increase of a known quantity of water, following combustion of the sample solid. The experiences carried out at the Italian Water Research Institute allowed to set up a standardised methodology consisting in preparing a table of dry sludge sample and then in estimating the temperature difference. Operating procedure is in accordance with ASTM D3286.

Vesilind (45) in a recent paper draws attention between a higher heating value (HHV) obtained by the use of the bomb calorimeter and a lower heat value (LHV) available for sludge incineration, the difference caused by the condensation of water vapor in the calorimeter (79.7 calories per gram of water formed from combustion).

Hiraoka et al (44) reported other two methods named thermogravimetric analysis (TG) and differential thermal analysis (DTA), very useful for classifying municipal refuse and sewage sludge. The TG analysis consists in the continuous determination of the weight of sample vs temperature. The DTA is based on the use of a calorimeter by which it is possible to observe the difference of the temperature produced by a standardised sample and by dried sludge. TG and DTA curves give, in general, various kinds of qualitative information about the characteristics features of thermal reaction process.

Table 5 : Physical properties - Identification of physical state

State	Definition
Liquid	Sludge will drain from a vessel through a calibrated orifice in a specific time. This is the
Plastic	The state in which material can undergo permanent deformation if the mechanical stress applied to it exceeds a certain threshold.
Solid state "with" shrinkage	This is a solid state in which the volume decreases as drying progresses, i.e. the medium is still saturated with water
Solid state "without" shrinking	No reduction in volume upon drying

Table 6 : Characterisation tests

State	Property	Test
Liquid	Viscosity	Flow rate through a specified viscometer
Plastic	Degree of plasticity	Pressure needed to extrude sludge through a specified orifice
	Mechanical stability	Casagrande shock apparatus
	Shear strength	Penetrometry test
	Adhesion	Force needed to extract specified cones from adhesion to plastic sludge
Solid state	Degree of shrinking	Volume reduction upon drying

9. Characterisation of sludges for handling and landfilling

Dewatered sludge water content is not sufficient to forecast mechanical properties of the material which determine the choice of the handling devices and the behaviour of the waste in a landfill. Since the general review made by Colin (30) on mechanical properties of sludges, recent progresses were made by using several geotechnical methods on raw dewatered sludge and sludge sampled in landfills (31, 32). The measured parameters are the bearing capacity (kN/m^2), the shear strength by using an apparatus developed for rapid determination of shear strength in cohesive soils (Vane), and the cone shear strength. There is no linear relationship between the shear strength and the dry matter content of the sludge.

The influence of the age of the sludge in the landfill on the geotechnical properties is not clear: in two cases there was no significant changes along the time. whereas in the third case a significant increase of shear strength is thought to be due to the low thickness after natural dewatering.

Several geotechnical methods were compared and discussed in a COST 681 workshop (33) held in 1987.

Konodromos (34) compared on sludges and domestic wastes several German standardised methods (DIN standards n°18 136, 52010, 51067, 4094, 4096).

10. Disposal of sludge to agriculture

Agricultural use of sewage sludge is still widely practiced in Europe and other parts of the world, particularly USA. Agricultural reuse is regulated by the 86/278 EEC directive or stricter national regulation. In USA the rule 503 applies.

These rules limit the maximum permissible concentration of heavy metals in sludge and soil and therefore standard methods are required for their determination.
Apart from heavy metals other parameters are required such as nutrients (nitrogen, phosphorus, potassium) and organic micropollutants (PCBs). In the framework of COST 681 several inter-laboratory comparisons led to following recommends regarding the analysis of total metal content in sludge.

The use of standard reference material for internal quality control would be a permanent requirement.
The use of a limited number of simple and reliable digestion methods are preferable instead of a large number of individual methods.
Among the methods recommended for use, digestion with aqua-regia proved the most suitable and reliable method and should be preferred to dry ashing.

As regards PCBs there has been a rapid development of analytical techniques using gas-chromatography with sensitive detection systems. The results of inter-laboratory comparison were published in a special report by EEC in 1985.

REFERENCES

(1) COMMISSION OF THE EUROPEAN COMMUNITIES. COST 68 "Sewage Sludge Processing". Final Report of the Management Committee, 1975.
(2) PUOLANNE, J. Sludge Production Rates. In "Processing and Use of Sewage Sludge" (Eds P. L' Hermite and H. Ott). D. Reidel Publishing Co. Dordrecht, 1984.
(3) NEWMAN, P. J., BOWDEN, A. V. and BRUCE, A. M., Production, Treatment and Handling of Sewage Sludge. In Sewage Sludge Treatment and Use" (Eds A. H. Dirkzwager and P. L' Hermite). Elsevier Applied Science. London, 1989.
(4) BRUCE, A. M., A Note on the Measurement of Sewage Sludge Production. In Treatment and Use of Sewage Sludge. COST 68 bis III. Technical Annexes. Commission of the European Communities, 1981.
(5) HOYLAND, G. and RONALD, D., Biological Filtration of Finely-screened Sewage. Water research Report TR 198. Water Research Centre, Medmanham, 1984.
(6) BRUCE, A. M., Compendium of Methods Used for Characterization of Sewage Sludges in relation to design and operation of Sludge Treatment Processes. "In Final Report (II Scientific Report) Concerted Action. Treatment and Use of Sewage Sludge". Commission of the European Communities. Brussels, 1983.
(7) HUAGAN, B. E. and MININNI, G. Characterization of Sewage Sludges. In "Characterization, Treatment and Use of Sewage Sludge" (Eds P. L' Hermite and H. Ott). D. Reidel Publishing Co. Dordrecht, 1981.
(8) HEIDE, H. W. and KAMPFE, R., Selection and use of polyelectrolytes in dewatering of sludge with belt presses in "Methods of Characterization of Sewage Sludge" (Eds T. J. Casey, P. L' Hermite and P. J. Newman). D. Reidel Publishing Co. Dordrecht, 1984.
(9) SPINOSA, L. and MININNI, G., Assessment of Sludge Centrifugability in "Methods of Characterization of Sewage Sludge" (Eds T. J. Casey, P. L' Hermite and P. J. Newman). D. Reidel Publishing Co. Dordrecht, 1984.
(10) CAMPBEL, H. W. and CRESCUOLO, P. J., Assessment of Sludge Conditionability using Rheological Properties. In "Methods of Characterization of Sewage Sludge" (Eds T. J. Casey, P. L' Hermite and P. J. Newman). D. Reidel Publishing Co. Dordrecht, 1984.
(11) HOYLAND, G. and DAY, M., Design Procedures for Batch Consolidation of Sewage Sludges. In "Methods of Characterization of Sewage Sludge" (Eds T. J. Casey, P. L' Hermite and P. J. Newman). D. Reidel Publishing Co. Dordrecht, 1984.
(12) COLIN, F., Characterization of the Physical State of Sludge. In "Methods of Characterization of Sewage Sludge" (Eds T. J. Casey, P. L' Hermite and P. J. Newman). D. Reidel Publishing Co. Dordrecht, 1984.
(13) COLIN, F., Experiences in the Laboratory with Columns and Lysimeters for the Study of Sludge Landfilling. In "Recent Developments in Sewage Sludge Processing" (Eds F. Colin, P. J. Newman and Y. J. Puolanne). Elsevier Applied Science. London and New York, 1991.
(14) GEUZENS, P. and DIELTJENS, W, Mechanical Strength Determinations of Cohesive Sludges. A Belgium Project on Sludge Consistency. In "Recent Developments in Sewage Sludge Processing" (Eds F. Colin, P. J. Newman and Y. J. Puolanne). Elsevier Applied Science. London and New York, 1991.

(15) STUCKIE, D. C., Assessment of the anaerobic Digestibility of Sludge. In "Methods of Characterization of Sewage Sludge" (Eds T. J. Casey, P. L' Hermite and P. J. Newman). D. Reidel Publishing Co. Dordrecht, 1984.
(16) BRUCE, A. M., Assessment of sludge Stability. In "Methods of Characterization of Sewage Sludge" (Eds T. J. Casey, P. L' Hermite and P. J. Newman). D. Reidel Publishing Co. Dordrecht., 1984.
(17) LOTITO, V., SPINOSA, L. and SANTORI, M., Influence of Digestion on Sludge Processing" In "Recent Developments in Sewage Sludge Processing" (Eds F. Colin, P. J. Newman and Y. J. Puolanne). Elsevier Applied Science. London and New York, 1991.
(18) FROST, R. C., Inter-relation between sludge Characteristics. In "Methods of Characterization of Sewage Sludge" (Eds T. J. Casey, P. L' Hermite and P. J. Newman). D. Reidel Publishing Co. Dordrecht, 1984.
(19) SPRINGER, A., et al, Innovative Procedure for automatic measurement of specific resistance to filtration and Electrostatic change Tappi Journal , Vol.77 n8 p.121, 1994.
(20) LEE, D. J. AND HSU, Y. H., Use of capillary suction apparatus for estimating average specific resistance to filtration cake. J. Chemical Technology and Biotechnology, Vol.59, n1, p 45, 1994.
(21) CHEN, C. W., LIN, W. W., LEE, D. Capillary suction time as a measurement of sludge dewaterability. Water Sci. Technol. Vol.34, n3/4 p.443, 1996.
(22) HERWIJN AREND J. M., et al. Determination of Specific cake resistance with a new capillary suction Time apparatus. Industrial and Engineering chemistry research, Vol.34, n4, p.130, 1995.
(23) CHANG, Y.C, NEETHLING, J. B. Viability of Anaerobic Digester sludge. J. of Environ. Engine., Vol 116, n 2, p 330, 1990.
(24) DOLD, P. L. et al. Assay for determination of A-amylase activity in activated sludge mixed bacterial communities. Environmental Technology, Vol.16, n2, p.181, 1995.
(25) WIESMAN, Z. and GRAFI, G. Indo-3-acetic acid (IAA) and cytokinin like activity in municipal excess activated sludge. Soil Science and plant nutrition, Vol.40 n1, p.117, 1994.
(26) CACOSSA, K. F. Calibration of a compressive gravity thickening model from a single batch settling curve. Water Sci. Technol., Vol.30, n8, p.107, 1994.
(27) PUN, K. C., et al. Characterization of sewage sludge and toxicity evaluation with Microalgae. Marine pollution Bulletin, Vol 31, n12, p. 394, 1995.
(28) SPANJERS, H., et al. Respinometry as a tool of rapid characterizing of wastewater and activated sludge. Water Sci. and Technol. Vol.31, n5, p.105, 1995.
(29) LESCHBER, R. Sludge characterization and standardization methods, development, status, trends. Water Sci. and Technol., Vol. 30, n8, p.81, 1994.
(30) BOCZAR, B. A., et al. Characteristics of enzyme activity in activated sludge using rapid analyses for specific hydrolases. Water Env. Research, Vol.64, n6, p.792, 1992.
(31) CHANG-WON, K, et al. Int-dehydrogenase activity test for assessing chlorine and H_2O_2 inhibition of filamentous pure culture and activated sludge. Water Research, Vol.28, n5 p.117, 1994.
(32) Monitoring biochemical activity during sewage sludge composting Biology and fertility soils (1993), Vol.16, n2 p.141, 1993.
(33) CHUNG Y-C. Microbial Activity measurement for anaerobic sludge digesters, Journal, Vol.61, n3, p.343, 1989.

(34) HOYGAARD, B. J. et al. On the stability of activated sludge flocs with implication to dewatering. Water Research, Vol.26, n12, p.1957, 1992.
(35) LOCKYEAR, D. F. and WHITE, M. J. D. The WRC thickenability test using a low speed centifuge. WRC Technical report 118, 1979.
(36) COLIN, F., GRAPIN, G., Bernard, C., GRANDJON, J. Utilisation d' une centrifugeuse à effet stroboscopique pour la caracterisation des boues et la prevision de leur epaississement gravitaire. Tribune du Cebedeau Vol. 40, n522, pp.27-36, 1987.
(37) COLIN, F., GRAPIN, G., Bernard, C., GRANDJON, J., Contribution des essais d' epaississement a la caracterisation des boues residuaires. Tribune du Cebedeau Vol.40 n523-524 pp.17-28, 1987.
(38) Development of settling flux curves using SVI. An addentum Water Envi. Res. Vol.67, n5 p.872, 1995.
(39) AGER M. C. On the origin of specific resistance of to filtration. Water Science and Technol., Vol 2, n12, p.58, 1990.
(40) UK Department of the Environment, Amenability of sewage sludge to anaerobic digestion HMSO, London, 1978.
(41) Sludge cinditioning controller uses on-line measurements of rheological properties. Plus sidebar on fuel oil from sludge. Water eng. & management. Vol.137, n6, p.31, 1990.
(42) MILENKO, R., Milan, D., FARKAS, A. P. Measurement of respiration of activated sludge. Water Research, Vol.22, n11, p.1405, 1988.
(43) NAGANAWA, T., KYUMA, K., YAMAMOTO, H. Measurement of soil respiration in the field: influence of temperature, moisture level, and application of sewage sludge compost and agro-chemicals. Soil science and plant nutrition, Vol.35, n4, p.509, 1989.
(44) Managing Sludge Through In-Vessel Composting, Water Eng. & management, Vol 138, n12, p.21, 1991.
(45) An Approach to the Control of return Sludge Pumping. Water Eng. & management Vol 143, n2, p.30, 1996.
(46) HIRAOKA, M., TAKEDA, N., SAKAI, S. "On thermal properties of conbustible solid waste. Asian Environment, Third Quarter, p.34, 1983.
(47) VESILING, A., and RANSEY, T. B., "Effect of drying temperature on the fuel value of wastewater sludge" "Sludge network" ,1996.
(48) DIN, Determintion of the amenability to anaerobic digestion. German Standard DIN 38414 Part 8, 1985.
(49) SOLLFRANK, U. and EUJER, W. Simultaneous determination of oxygen uptake rate and oxygen tranfer coefficient in activated sludge systems by an on-line method Water Research, Vol 24, n 6, p. 725, 1990.

ON-LINE MEASUREMENT IN WASTE WATER TREATMENT PLANTS: SENSOR DEVELOPMENT AND ASSESSMENT OF COMPARABILITY OF ON-LINE SENSORS.

H. S. JACOBSEN and A. LYNGGAARD JENSEN, VKI - Department of Monitoring and Information Technology, Science Park Aarhus, DK-8000 Aarhus C., Denmark.

1. Introduction

Overviews of commercially available sensors (e.g. analysers) for water treatment plants[1] show that many of these require sampling and filtration. Although these sensors are improving[2-3], they still require a good deal of maintenance[4-7] and the installation costs can be considerable. Furthermore, long response times (10 - 25 minutes) may cause difficulties in incorporating them in standard feed-back process control loops.

Thus the need for new principles for sensors for real-time on-line process control seems obvious and some of the major demands to the development work are and have been to design sensors, which can be placed directly in activated sludge (no sampling and filtration) and measure within appropriate ranges with a fast response time (less than 5 minutes) and continue to do that with maintenance only once a month, or more.

Several well known measuring principles can be used to comply with these demands, if the principles are used together with technology now available (such as possibilities for fast and large computations, optical standard components and micro-machined designs). Some of these principles include:

Colorimetry combined with semi-micro mechanics and membrane technology.
Absorbance using photo diode-arrays for spectral information.
Fluorescence with multi-wavelength excitation and scanning for emissions.
Virtual sensors (or software sensors) based on calculations on signals from reliable sensors or installed equipment.

Once the sensors have been developed, the next question to arise is: How is the reliability and quality of measurements obtained from the sensors secured and how is the comparability of different sensors improved.

At the present time neither validation schemes nor standards exist for user needs concerning the in situ on-line measurements of water quality determinands and in consequence, the instrument producers and suppliers set their own specifications, based upon the technology that they possess.

These problems are the basis of a recently started project, financed by the SMT programme, the ETACS (European Testing and Assessment of Comparability of on-line Sensors/analysers) project. The objective of the project is to initialise a process, which will establish a validation scheme, which will have the form of a test protocol, that can be used in two situations :

- in the laboratory for documentation of the instrumentation according to the performance claimed by the producers.
- in situ during operation for documentation of the reliability of results obtained by the users.

The content of a test protocol, which can be the technical basis for further work in European and international standardisation organisations, should be based on the typical performance characteristics of in situ on-line sensors, which include :

- Selectivity & specificity
- Range
- Linearity
- Sensitivity
- Limit of detection
- Limit of quantification
- Ruggedness
- Accuracy
- Precision, repeatability & reproducibility
- Response time, dead time & rise time
- Effective operation time (up time)

The project will suggest precise definitions of these performance characteristics and test the practical use of them, both in the laboratory and in the field.

The paper will concentrate on the two aspects mentioned above. a) Description of a sensor developed using the first of the four measuring principles, although brief description will be given for the other three mentioned principles. b) Further description of the need for a test protocol, to secure the quality and reliability of sensor measurement and to improve the comparability of different sensors/analysers.

2. Sensor development

2. 1. Colorimetric sensors

The sensors developed using colorimetric principles, in combination with semi-micro mechanics and membrane technologies, include ammonium, nitrate and phosphate sensors and is defining a concept which now is called μCFA (Micro Continuous Flow Analysis).

2. 1. 1. Sensor design

Basically the principles used are the same for all three kinds of sensors. Fig. 1 shows a diagram of how the sensor is working and how the different parts of the sensor have been built into the prototype construction. Reagents and clean water are stored in bags (1) and transported around the sensor with pumps mounted on a transportation manifold (2). A carrier (clean water) is pumped to the membrane device (3) and flows in a channel on one side of the membrane. The other side is exposed to the measuring media and the carrier is then enriched with ions passing through the membrane. The enriched carrier is mixed with reagents and pumped to the reaction area (4) and further on to the photometer for detection (5). Both the reaction area and the photometer are heated to a constant temperature. From the photometer all liquids are pumped to a waste bag (6). All functions in the sensor are supervised and controlled by a microcomputer (7).

Figure 1: Design and construction of the sensor.

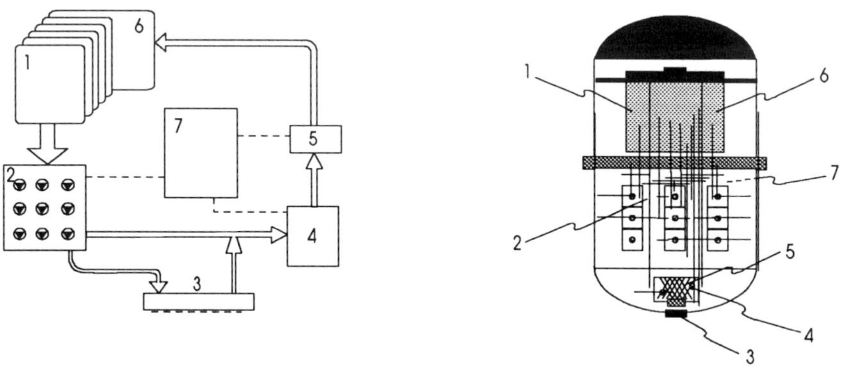

Transport manifold : All transportation of liquids inside the sensor takes place in channels with a width of 0.3 to 1 mm milled directly into the manifold material. Pumps delivering 3 μl/turn, are integrated with the manifold and deliver carrier and reagents from the cooled storage bags with a speed of 5 rpm, resulting in a total consumption of liquids within the sensor of less than 3 litres/month. The length of the channels (also on the analysis manifold) have been minimised in order to obtain the smallest possible volume of flowing liquids, giving the fastest response time as possible.

Membrane device : The membrane functions as an ion sieve with practically no mass transportation of water. This means that the sensors with respect to liquids can be regarded as a closed loop system with all used liquids collected in the waste bag. As the carrier is flowing continuously on one side of the membrane and the other side is in direct contact with the measuring media, the sensors can deliver a true continuous measurement, without the need for any further sample pretreatment (pumps, cross flow filters, etc.). The MW cut-off of the membrane is approximately 100 Daltons and with an approximate thickness of 50 μm, which is the same as for many Dissolved Oxygen sensors. The lifetime of the membrane is tested to be well over 1 month.

Analysis manifold : On the analysis manifold, reagents and carrier are mixed, starting a colorimetric reaction and the resulting colour is measured by the sensor photometer. Transport inside the manifold is performed as in the transport manifold and these two manifold and the membrane device define the μCFA system.
The analysis manifold, together with the sensor photometer (or the wavelength of the photometer), gives the sensor its identity and three different manifolds have been developed for well-known colorimetric reactions: Indophenol Blue method for ammonium determination, hydrazine reduction followed by formation of a Diazo complex for nitrate determination and Molybdenum blue method for phosphate determination. However, the high demand for short response times makes it impossible to allow the reactions to come to an end and this problem has been solved by design of a system with very high repeatability enabling the photometer detection to take place at the exact same time during the colorimetric reaction.

Calibration and measurement : Beside the normal operation mode, the sensors are equipped with an auto-calibration and a cleaning routine, using standard solutions of the ion to be measured and proper cleaning agents. Both the standards and cleaning agents are stored in bags together with the reagents and each bag, for measuring, calibration and cleaning, has its own pump, controlled by the microcomputer. This enables the sensor to operate in different modes (measuring mode, calibration mode, etc.), according to preset criteria. The standards can be used in two different situations: calibration of the chemical system, or determination of the membrane efficiency. Both calibration routines and the cleaning routine are performed with the sensor placed in the measuring media.

For low concentrations of the actual ion to be measured transmission follows Lambert-Beers law. However, for high concentrations this law does not apply and therefore the dynamic range is extended using a calibration involving three standard references and a non-linear relationship between concentration and absorbance for higher concentrations.

$$C^* = \mathrm{Log}(Tr/a)/b \quad \text{(for low concentrations) and} \quad C^* = c/(Tr-d) \quad \text{(for high concentrations)} \quad (1)$$

where C^* is the resulting concentration measured by the chemical system, Tr is the measured transmission and a, b, c and d are the constants calculated from the measured transmission with low and medium reference concentrations, respectively medium and high reference concentrations.

As the membrane efficiency M_{eff}, can be lower than 100% and can vary from one membrane to another, it is essential that the sensor automatically determines this efficiency, in order to be able to calculate the correct concentration of the ion in the measuring media, defined as $C = C^*/M_{eff}$. It can be shown that[8]:

$$M_{eff} = 1+[(C^*-C_s^*)/C_s] \quad (2)$$

where C_s is the concentration of the reference in the carrier before passage of the membrane and C_s^* is the concentration of the reference after passage of the membrane.

Signal conditioning: The raw analogue photometer signal is sampled and digitised with a frequency of 10 Hz and is then filtered for peaks caused by noise or air bubbles passing through the photometer. Finally a moving average is calculated with a time period of 12 seconds (which equals one turn of the pumps and 120 sampled values), in order to remove the effect of oscillation of the signal caused by the pumps. This averaged value is then used by equation (1) and finally divided by the membrane efficiency.

2. 1. 2. Laboratory tests

Laboratory tests were designed to give information regarding the sensor performance according to the definitions given in EN45001[9], of performance characteristics of an instrument. Calibration curves for each of the three sensor types were determined using standard solutions and each sensor, in turn, was placed in these standard solutions for an hour, logging every 5 seconds, giving 720 measurements for each sensor in each standard solution. Response times for the sensors were determined using standard solutions, as the sensors were moved rapidly from a blank solution to a standard solution and were defined as the time it takes the sensor to reach 90% of a new concentration, caused by a step change. Table 1 shows the result of the laboratory tests for some performance characteristics.

Table 1. Sensor performance according to laboratory tests.

	Unit	**NO_3^- - N**	**NH_4^+ - N**	**PO_4^{3-} - P**
Range	mg/l	0 - 10	0 - 25	0 - 10
Sensitivity	mg/l	0.05	0.05	0.025
Limit of quantification	mg/l	0.05	0.05	0.025
Limit of detection	mg/l	0.075	0.1	0.2
Accuracy*	mg/l	±(0.1 + 10%)	±(0.1 + 5%)	±(0.1 + 5%)
Repeatability*	mg/l	±(0.1 + 5%)	±(0.1 + 3%)	±(0.2 + 3%)
Response time	min.	3	2½	5

*percentage is given of the actual measured concentrations.

2. 1. 3. Wastewater Treatment Plant Tests

The sensor prototypes have been tested in a pilot scale waste water treatment plant operating with recirculation nitrogen removal and in a full scale biodenitro waste water treatment plant. The pilot scale plant was loaded with real wastewater and both treatment plants had extended nitrogen removal.

Fouling of the sensors: To investigate the fouling of the membrane devices, ten devices were placed in the pilot scale treatment plant for a period over half a year and laboratory analyses were carried out periodically on the carrier from each of the membrane devices and from the tank of the pilot scale treatment plant.

During the period the pilot scale treatment plant was operated as close to steady state as possible and the feed to the plant consisted of raw waste water. The mass transfer coefficient for the membranes was calculated from the experimental results[8] and it was found that the ion transport over the membrane was practically unaffected during the period. This result was confirmed by visual inspection of the membranes during the period, which showed no bacterial growth at all.

Sensor responses to the process in the pilot scale treatment plant: The sensors were installed in the aeration tank of the pilot scale treatment plant and several experiments performed over a longer time period with determination of process rates were conducted using the sensor responses and laboratory analyses on grab samples from the tank. The tank functioned as a batch container, due to stop of the inflow of feed during the experimental period.
All recorded sensor responses are in good agreement with the laboratory analyses and the sensor response times are very close to follow real time changes in the pilot scale treatment plant. Fig. 2 shows that the calibration and response of the sensors (ammonium sensor) do not change over several days, compared with laboratory analyses and that the pilot scale treatment plant is normally operated with a very low concentration of ammonium in the nitrification tank. The figure shows several experiments with adding of ammonium chloride (first and second peak) or stop/start of the aeration.

Figure 2. Ammonium measurement in the pilot scale treatment plant.

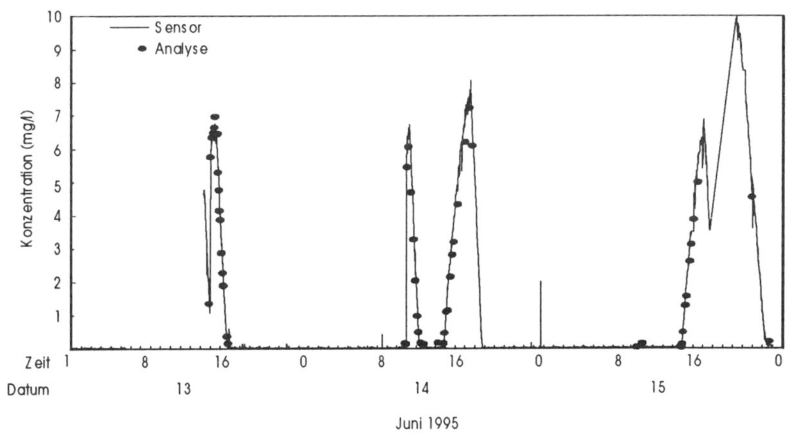

Sensor response to the process in the full scale biodenitro waste water treatment plant: Figure 3 shows the ammonium sensor response during three days of test, in the full scale waste water treatment plant (from Friday afternoon to Monday morning). Responses are as expected both the very high load (which is partly because a nearby slaughterhouse cleaned up for the weekend) and the low load during weekend. Autocalibration periods can easily be seen as periods of constant signal around midnight and it is obvious that calibration every night is not needed. Furthermore, it can easily be seen that the control scheme of the plant can be optimised, because aeration periods are clearly too long in the weekend and maybe too short during periods of high load.

Figure 3. Ammonium measurements in the full scale waste water treatment plant.

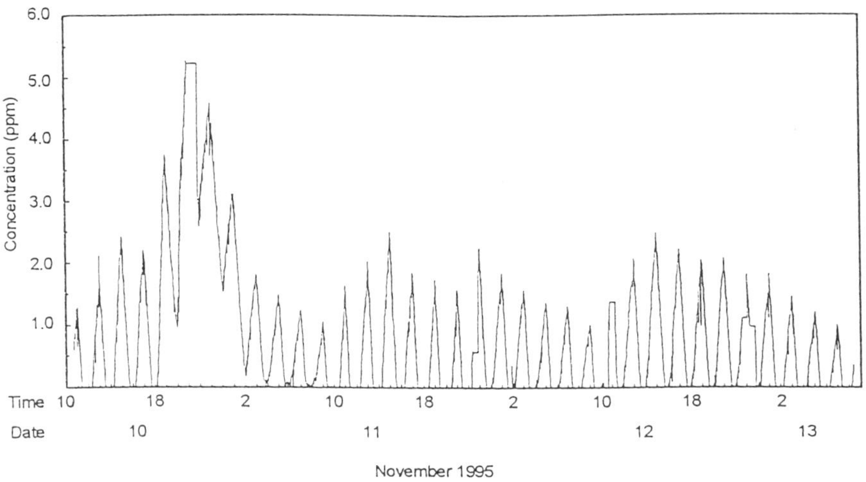

2. 1. 4. Conclusions

The tests of the prototypes of the sensors demonstrate that it is possible to design and construct real-time on-line in-situ sensors (without any pretreatment of samples and without any spill of chemicals (closed loop system)), with response times less than 5 minutes for ammonium, nitrate and phosphate. Furthermore, the sensors have an effective auto-calibration routine and no fouling of the membranes used is observed and finally maintenance periods of a month is possible, due to the low chemical consumption. The sensors have now been brought from prototypes to commercially available sensors and are now available on the market.

The μCFA and membrane technology developed for the prototype sensors are believed to have a large potential for further refinements, such as extension of the dynamic range and faster response times. Furthermore, it should be possible, without too much difficulty, to extend the types of ions to be measured by the sensors.

It has been demonstrated[10] that this type of sensors supports the use of set-point control by simple rule based control schemes and it is believed that there is a great potential for optimisation of treatment processes even with these simple control schemes.

2. 2. Absorbance, Fluorescence and Software Sensors

2. 2. 1. Absorbance sensors

The sensors presented in the previous section measure absorbance at one specific wavelength, specified by the developed colour. However, it is also possible to measure the absorbance spectrum of a sample by sending light through the sample and measure the absorbance at the different wavelength. As every constituent in the sample, which have molecular energy levels, which are able to absorb the light used will absorb the light with different absorbances, due to the specific energy levels for the specific constituents, complex samples can give very complex spectra.

Absorbance measurements is usually obtained in the UV-VIS (Ultraviolet and Visible light) region of the spectrum. This region has a lower limit due to the strong absorbance of water itself below 200 nm. Nitrate is a good example of a constituent that has a specific absorbance (at 205 nm), but in waste water samples this absorbance is very often hidden by strong absorbances from organic substances. Nevertheless, nitrate sensors, based on the absorbance at 205 nm, are available on the market, all of them trying to compensate for the interference from organic substances.

Figure 4. Absorbance spectra in the UV region of samples from a waste water treatment plant.

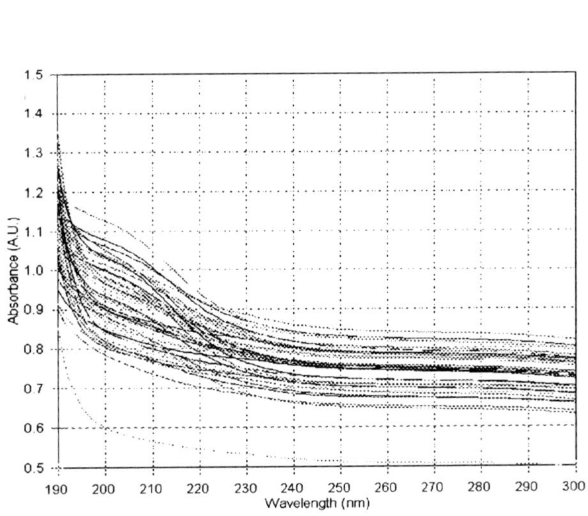

Figure 4 shows absorbance spectra of samples of activated sludge at different times during the cycle of a biodenitro waste water treatment plant and as can be seen from the figure, these samples are all very complicated. However, the samples can be analysed for different substances, using multivariate calibration and this calibration can be used to give the concentrations of the substances, of spectra recorded at a later time. This principle is clearly demonstrated for nitrate by Karlsson et at.[11], who also showed that it would be possible to measure total-P, total-N, ammonium and iron simultaneously with the nitrate. This methodology is also known as chemometrics and together with the development of micro-scale spectrophotometers (diode arrays) it will be possible to develop small integrated multiparameter absorbance sensors.

2. 2. 2. Fluorescence sensors

As described above, absorbance is caused by substances absorbing light. The substances, which are now in a higher and more unstable energy state, release the energy surplus. This release is normally done by molecular processes like rotation, vibration or transfer of energy to other molecules. However, in some substances this release can happen as emission of light and this is known as fluorescence. The wavelength of the fluorescent light is specific for a given substance and will always be at a longer wavelength than the light absorbed, because some of the energy surplus always will be released as vibration and rotation.
A typical design of a single channel fluorimeter (sensor based on fluorescence) is described by Nørgaard et al.[12] This fluorimeter measures the NADH concentration by the fluorescence at 460 nm (excitation at 340 nm), thereby giving a measure of the biological activity, as the corresponding oxidised state, NAD^+, is not excitated at 340 nm. This fact has been known for several years in the fermentation industry.

Furthermore, it can be shown[13] that the NADH fluorescence can be used to detect the end of the denitrification in an alternating activated sludge plant and that some correlation exists between fluorimeter output and the peak oxygen uptake rate.

As with absorbance, also fluorescence can be measured as a spectrum. This spectrum has a 3-dimensional structure and is the fluorescence as a function of both excitation and emission. Until today only very little work has been done in the field of multi-channel fluorescence and chemocmetrics[14] and almost nothing on waste water samples.

2. 2. 3. Software sensors

Software sensors, or virtual sensors, is a term used for "signals" from calculations based on signals from reliable, available sensors in combination with other signals like on/off indications and time counters.

As an example of a software sensor, consider the rate of change of the dissolved oxygen concentration, $d(C_{ox})/dt$. This rate change can be calculated from the oxygen concentration measurements, C_{ox}, done at time n and n-1

$$d(C_{ox})/dt = (C_{ox,n}-C_{ox,n-1})/(t_n-t_{n-1}) \quad (3)$$

Furthermore, the rate change can also be expressed in terms of respiration

$D(C_{ox})/dt = K_l a(C_{ox,m}-C_{ox})-OUR$ (4)

As the oxygen transfer coefficient, $K_l a$, can be written as a new constant multiplied with the airflow, Q_{air}, to the aeration tank and as the oxygen saturation concentration, $C_{ox,m}$, can be calculated as

$C_{ox,m} = C_{ox,20}\ 1.02^{(20-T)}$ (5)

then the oxygen uptake rate, OUR, can be written as

$OUR = K \cdot Q_{air}(9.07 \cdot 1.02^{(20-T)}-C_{ox})-d(C_{ox})/dt$ (6)

K is determined by setting $Q_{air} = 0$, ($OUR = d(C_{ox})/dt$) and following this by trial and error setting a Q_{air} giving a constant oxygen concentration, ($d(C_{ox})/dt = 0$). Hereafter the OUR can be calculated from measurements done by the oxygen sensor, the temperature and the airflow.

Other kinds of software sensors are for example filters, to be used in real time in order to avoid outliers in the signal from the real sensors or different types of real time quality control/sensor validation procedures. These software sensors are in fact sensors on the sensors and can be useful to follow.

3. Assessment of comparability of sensors, the ETACS[15] project

In order to support the protection of the environment, there is an urgent need for improvement of comparability, reliability and quality of measurements obtained from in situ on-line sensors/analysers, used to determine the composition of and flow of effluents from waste water treatment plants and industries.

Therefore a new project, the ETACS[15] (European Testing and Assessment of Comparability of on-line Sensors/analysers) project has recently been initiated. The purpose of the project is to carry out a pre-normative work on standardised validation of methods for in situ on-line measurements of water quality determinands, such as ammonium, nitrate, phosphate, BOD, COD and TOC. The key objectives of the project are:

The development of a test protocol for validation of the performance of in situ on-line sensors/analysers and how these can be compared.
Practical testing of the test protocol to access its applicability to provide experience to form a final test protocol.
Widespread acceptance of the test protocol by producers/suppliers, users and relevant authorities, to assist its early adoption as an agreed European standard.

These objectives will secure that in situ on-line water quality measurements will be of sufficiently known and documented quality, in order to be used for real time process control, regulatory monitoring and harmonised environmental databases.

At the present time neither validation schemes nor standards exist concerning the in situ on-line measurement of water quality determinands and in consequence the producers and suppliers set their own specifications, based upon the technology that they possess. This approach to measurements is very unsatisfactory, since it provides no basis for users to either specify or compare performance of the sensors. In addition to this, producer supported data are often based on laboratory tests, carried out off-line and therefore not carried out on genuine on-line samples.

Therefore the ETACS project shall initialise a process, which will establish a validation scheme, which will have the form of a test protocol and which eventually can form the basis of a system for accredited certification of in situ on-line sensors/analysers, enabling reliable and comparable measurements. The validation scheme can be used in two situations :

- in the laboratory for documentation of the instrumentation according to the performance claimed by the producers.
- in situ during operation for documentation of the reliability of results obtained by the users.

The test protocol shall be independent of both specific in situ on-line sensors/analysers and specific determinands and during the project period it will be tested and documented as the technical basis for further work in the standardisation organisations.

The content of the test protocol shall be based upon the typical performance characteristics of in situ on-line sensors/analysers and include:

Selectivity & specificity :

- Range.
- Linearity.
- Sensitivity.
- Limit of detection.
- Limit of quantification.
- Ruggedness.
- Accuracy.
- Precision, repeatability & reproducibility.
- Response time, dead time & rise time.
- Effective operation time (up-time).

The project will suggest precise definitions of these performance parameters and test the practical use of them both in the laboratory and in the field.

The project will fall into three phases :

Phase 1

The development of a test protocol for validation of the performance of in situ on-line sensors/analysers and how these can be compared.
Collection of information about existing standards and certification- /accreditation arrangements in Europe.
Information to the European Standard Organisations - with the possibility to form groups for discussion and information dissemination.
Detailed description of the procedures and design of methods to be used for laboratory tests and field test, based on a review of existing work and experience.
Identification of determinands and producers/suppliers of sensors/analysers, to be considered to be covered by the test protocol.
Identification of at least three different types of on-line sensors/analysers, which will be the best representatives for testing the developed procedures and methods.
Preparations of a draft test protocol for laboratory and field tests of on-line sensors/analysers.
Information to the European producers/suppliers and laboratories about the work and the possibilities for participation in phase 3 of the project.

Phase 2

Practical testing of the test protocol to access its applicability to provide experience to form a final test protocol.
Selection of determinands and sensor/analyser classes.
Purchase of sensors/analyser and development of the logistics, "test bench facilities" and standard samples.
Commissioning of the chosen types of sensors/analysers.
Laboratory and field test of the draft test protocol on at least two different types of on-line sensors/analysers. Test is performed by the project participants.
Review of the protocol with revision of the procedures and methods, according to the experience obtained after both laboratory testing and field testing.
Meetings with a group of European Standard Organisations.
Production of a summary report, the final test protocol and other information to be used in phase 3 of the project.

Phase 3

Widespread acceptance of the test protocol by producers/suppliers, users and relevant authorities, to assist its early adoption as an agreed European standard.
Extension of the number of laboratories and offer European producers/suppliers of on-line sensors/analysers the opportunity to have there product tested at these laboratories, according to the protocol.
Formation of a network of laboratories and producers/suppliers, in order to be able to collect and disseminate the results and experience obtained.

Organise a series of workshops for end users in different countries, with presentations performed by participants, laboratories and producers/suppliers and with participation of different standardisation organisations.
Production and dissemination of a final report to all interested partners.

Currently we are in the beginning of phase 1 and collecting information about the current state of art, regarding existing standards of the above mentioned performance characteristics. Two searches have been conducted in all major databases, world-wide and the result showed that there is no standard available at all, covering the above mentioned performance characteristics, sensors and water/wastewater and that only a few of them have been defined.

This result confirms the basis of the ETACS project, that is to produce a test protocol that can be used to secure the quality and reliability of sensor measurements and used to improve to comparability of different sensors. This test protocol shall form the basis of further work in international standardisation organisations.

4. Acknowledgement

The sensor development has been done by Danfoss A/S in collaboration with VKI and other partners.
The ETACS project is identified as project 3256 - ETACS, founded by the SMT programme and managed by DGXII, for which is acknowledged.

REFERENCES.

1. A. Lynggaard-Jensen. Status for On-line Sensors and Automated Operation of Wastewater Treatment Plants. In: Matti Valve (ed) Nitrogen Removal from Municipal Wastewater. Nordic council of Ministers, Copenhagen,174-186 (1995).
2. H. Wacheux, S. Da Silva, J. Lesavre. Inventory and Assessment of Automated Nitrate Analysers for Urban Sewage Works. Wat. Sci. Tech. 28, 11-12; 489-498 (1993).
3. H. Wacheux, J. L. Million, C. Guillo and E. Alves. NH_4^+ Automatic Analysers for Wastewater Treatment Plant: Evaluation Test at laboratory and Field Level. .Wat. Sci. Tech. 33, 1; 193-201 (1996).
4. S. Schlegel and P. Baumann. Requirements with Respect to On-line Analysers for N and P. Wat. Sci. Tech. 33, 1; 139-146 (1996).
5. H. A. Thomsen and K. Kisbye. N and P On-line Meters: Requirements, Maintenance and Stability. Wat. Sci. Tech. 33, 1; 147-157 (1996)
6. J. Londong and P. Wachtl. Six Years of Practical Experience with the Operation of On-line Analysers. Wat. Sci. Tech. 33, 1; 159-164 (1996).
7. U. Nyberg, B. Andersson and H. Aspegren. Experiences with On-line Measurement at a Wastewater Treatment Plant for Extended Nitrogen Removal. Wat. Sci. Tech. 33, 1; 175-182 (1996).

8. A. Lynggaard-Jensen, N. H. Eisum, I. Rasmussen, H. S. Jacobsen and T. Stenstrøm.Description and Test of a new Generation of Nutrient Sensors. Wat. Sci. Tech. 33, 1; 25-35 (1996).
9. EN45001. General Criteria for the Operation of Testing Laboratories. Explanatory Guide, section 15: Validation (1992).
10. P. Balslev, A. Lynggaard-Jensen and C. Nickelsen. Nutrient Sensor Based Real Time On-line Process Control of a Waste Water Treatment Plant using Recirculation. Wat. Sci. Tech. 33, 1; 183-192 (1996).
11. M. Karlsson, B. Karlberg and R.J.O. Olsson. Determination of Nitrate in Municipal Waste Water by UV Spectroscopy. Anal. Chim. Acta. 312; 107-113 (1995).
12. P. Nørgaard, K. Helmo and E.W. Sørensen. Treatment Process for Nitrogen Removal Controlled by NADH. Vand og Jord 3, 3; 125-129 (1996), (in Danish).
13. S. Isaacs and M. Henze. Fluorescence Monitoring of an Alternating Activated Sludge Process. Wat. Sci. Tech. 30, 4; 229-238 (1994).
14. P. Nørgaard. A Multivariate Chemometric Approach to Fluorescence Spectroscopy. Talanta 42; 1305-1324 (1995).
15. European Testing and Assessment of Comparability of on-line Sensors/Analysers. Project 3256 - ETACS. Part of the Standards, Measurement & Testing programme, Managed by DGXII.

FIBRE-OPTIC CHEMICAL SENSORS: FROM MOLECULAR ENGINEERING TO ENVIRONMENTAL ANALYTICAL CHEMISTRY IN THE FIELD

G. ORELLANA and D. GARCÍA-FRESNADILLO, Laboratory of Applied Photochemistry, Dept. Organic Chemistry, Faculty of Chemistry, Universidad Complutense de Madrid, E-28040 Madrid, Spain.
M. D. MARAZUELA, M. C. MORENO-BONDI, Dept. Analytical Chemistry, Faculty of Chemistry, Universidad Complutense de Madrid, E-28040 Madrid, Spain;
J. DELGADO and J. M. SICILIA, Grupo INTERLAB, María Tubau 4 - 3ºA1, E-28050 Madrid, Spain

1. Abstract

Fibre-optic chemical sensors or optodes can be an attractive alternative to current electrochemical or electronic devices in terms of monitoring stability, electrical hazards, size, target analytes, versatility, simplicity, ruggedness and cost. In order to show up their full potential, dedicated optoelectronic instrumentation and tailor-made optical indicators must be developed. We have recently combined efforts between industry and university in order to carry out a R+D project that spans from molecular design of novel luminescent Ru(II) polypyridyl dyes, to the probe engineering, fabrication, characterisation, validation and in-situ monitoring with optical devices based on waveguides and active sensing heads, capable of continuously monitoring relevant parameters in waters. Two examples of how the fibre-optic analysis of dissolved oxygen and carbon dioxide have been tackled by a suitable design of the indicator molecules, are discussed. Results on oxygen monitoring with optodes in the biological treatment pool of a wastewater sanitation plant are presented as well.

2. Introduction

Continuous sensing of relevant chemical determinands is a matter of growing interest in environmental monitoring. This is due to the fact of natural variations of their concentration (daily, seasonal, locational,...), as well as human-induced changes such as those caused by industrial pollution, domestic waste, transportation media, etc. It is nowadays recognized that prevention activities, e.g. improvement of industrial procedures or advanced monitoring, are far more cost-effective than remediation procedures to be used once the damage has been inflicted. It is not surprising, therefore, the current striving of the industrialized nations to deploy all kind of environmental sensors in the atmosphere and the hydrosphere.

In connection to the new technologies for protection of the environment, the european parliament and the EU council underline in their Decision no. 1110/94/EC the necessity of focusing the efforts in RTD on the instrumental technologies required for environmental observation, surveillance, and research. "This implies, particularly, a contribution to the development of technologies for observation of the Earth, [...], including herein sensors, observation technologies and surveillance of different biosphere behaviours, as well as a main development of the required analytical technologies for the different matrices. [...]. It is also to be included here all those instruments related to prevention of natural catastrophes and to pollution control of industrial installations."

Many techniques have been developed so far in response to specific regulatory or public safety requirements. The sensitivity, specificity and the versatility of optical techniques of analysis well established in the laboratory environment, make spectroscopy a very popular method on which to base continuous monitoring systems. Unfortunately, the features that make spectroscopy so desirable have, to date, made it costly and complex compared to the largely unexpensive mecanical and electrical sensors; the situation, however, may change in the nearest future.

A novel type of device for monitorization or quantitation of chemical parameters (ions, gases, organic chemicals,...) via light guides, has evolved recently from the fruitful combination of (apparently) disparate fields such as organic and analytical chemistry, photochemistry, polymer technology, telecommunication engineering, optoelectronics, physical optics and spectroscopy. Those devices have been called fibre optic chemical sensors (FOCS®) or "optodes" (from the greek "optical way"; the term "optrode"® has been used as well, because of the similarity of application to the classical electrodes).
Basically, a typical device of this kind (Figure 1) consists of a light source, the radiation of which (usually wavelength-selected) is launched into the core of a wave guide. The optical fibre carries the light to the sensitive tip, - the true "heart" of the device -, that contains an appropriate indicator (reagent layer) immobilized onto a polymer matrix.

Figure 1. Basic constituents of a fibre-optic chemical sensor (F: wavelength selectors; D: photon detector).

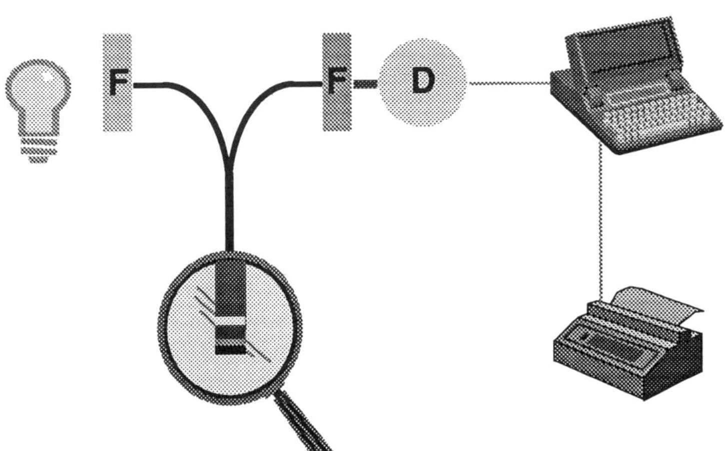

Some optical property (absorbance, reflectance, luminescence, reflectivity, refraction index,...) of this layer undergoes a measurable change upon (selective) interaction with the analyte of interest; in this way, the modified optical radiation is guided back via the fiber optics to a photon detector, the electrical signal of which is amplified and recorded analogical or digitally. A proper calibration allows measurement of the determinand concentration.

Optodes and practical optical sensing schemes have been developed successfully for quantitating gases (oxygen, carbon dioxide, ammonia, sulfur dioxide,...), cations (pH, potassium, calcium, copper, surfactants...), anions (chloride, iodide, cyanide, sulfide,...), organic pollutants (oils, chlorinated hydrocarbons,...), and physical parameters (turbidity, temperature,...) Our group has contributed to this field with optodes for pH, H_2S, iron, CO_2, oxygen, alcohols, humidity, and detergents, among other analytes. The advantages of using fibre optic sensors for chemical analysis and, particularly, environmental monitoring are noticeable: (a) as opposed to classical electrodes, the optical sensors require neither a reference signal (subject to fouling or drift), nor are subjected to electrical interferences (e.g. damage caused by electromagnetic fields of thunder storms); (b) they are easily miniaturisable to allow access to narrow monitoring sites; (c) they do not consume analyte during the measuring process; (d) it is possible to design devices capable of detecting chemical species which can not be detected by any current electrical or electronic sensors; (e) optical wires are able to transport a higher information density than their electrical counterparts; (f) smart optical sensors may be competitive in terms of operation and maintenance versatility, simplicity, ruggedness, and cost. If additionally we incorporate to the sensing membrane some compound of biological origin such as an enzime, an antibody or a carrier, and a proper optochemical transducer (e.g. pH, oxygen, hydrogen,...), we will end up with a so-called biosensor, an optical detector the specificity and sensitivity of which allow detection and quantification of minute amounts of selected chemical species (e.g. pesticides, explosives, poisons, ...).

The design and construction of optical sensors and biosensors should overcome, of course, a number of potential drawbacks: interference of ambient light, long-term stability due to (photo)decomposition or leaching of the supported indicator, necessity of a mass transfer step in order to reach a stable response, the observance of a limited response interval in many cases (e.g. sensors of sigmoidal response), and the still limited number of commercial components specifically designed for this application. Actually, the last 15 years have wittnessed the birth, infancy, and the arrival to the adult age of fiber-optic chemical sensors in the basic research of academia and industry, yet the gap to marketing and practical applications of such devices has to be filled up. It is a challenge to researchers and technologists in this field to be able to minimise or suppress such limitations in order to reach maturity for optodes.

We are currently engaged in an ambitious RTD project that spans from molecular design of novel optical indicators, to engineering, manufacturing, characterization, validation and in situ monitoring with optical devices based on fibre optics and "active" sensing heads (i.e., those containing a chemical indicator immobilised onto a polymer support and placed at the end of a light guide), capable of monitoring continuously the concentration of species that determine the quality of natural, industrial, irrigation, and drinking waters.
In order to perform in situ environmental monitoring, it is advisable to design small, battery-operated instrumentation based on solid-state technology (e.g. excitation with LED or laser

diodes, silicon photodiode detection, etc.). Typically, the detection system must be able to measure optical signals of a very low power (50 pW to 1 nW), with a signal-to-noise ratio better than 60 dB. Therefore, it is necessary to optimise the optoelectronic system in connection to the sensitive membranes of the device, in order to manufacture a low-cost portable instrument based on luminescence measurements (photoluminescent sensors have demonstrated usually to be more sensitive than absorbance-based devices, and more durable if based on emission decay determination or phase-sensitive detection rather than absolute luminescence intensity).

Moreover, it would be desirable that a single hardware might be used regardless the analyte being monitored, with the only change required being the sensitive tip of the optical fibre. To this end, it is our goal to design and prepare a single family of luminescent indicators capable of responding to the largest possible number of determinands relevant to water analysis (O_2, temperature, pH, CO_2, detergents, BOD and salinity).

Coordination compounds of ruthenium(II) and other transition metal ions with polyazaheterocyclic ligands are luminescent dyes with a high potential for chemical analysis. The spectroscopical and redox characteristics of such species can be finely "tuned" by a careful selection of the chelating ligands. They absorb strongly in the blue-green part of the electromagnetic spectrum and display significant luminescence in the red (600-750 nm; $10^{-3} < \Phi_{em} < 0.2$); these features fit nicely the performance of current emission and detection diodes. Their emission has a Stokes' shift larger than 150 nm, which facilitates separation of the scattered light that accompanies the measured luminescence, with simple coloured glasses. Moreover, the slight triplet character of the excited state of such complexes determines a rather long emission lifetime (0.03–6 μs); this feature enhances their sensitivity towards the analyte and makes possible to develop optical sensors using both fluorescence intensity and kinetics. In this way, based upon our experience in the synthesis, spectroscopy, photophysics and photochemical characterisation of luminescent complexes of ruthenium(II) with polyazaheterocyclic ligands, we have engineered such dyes to become the sought family of optical indicators to fabricate fibre-optic chemical sensors. Two cases in point from our research and development in this field will be presented in the following sections.

3. Experimental Section

The oxygen indicator tris(5-octadecamide-1,10-phenanthroline)ruthenium(II), abbreviated Ru(5-odap) (Figure 2), has been prepared from 5-odap (0.65 mmol) and ruthenium trichloride trihydrate (0.19 mmol, Janssen), by heating the solution at reflux in ethyleneglycol (20 mL, Panreac p.a.) under argon for 4 h. The orange complex ion is then precipitated upon addition of 10 mL of water and three drops of concentrated hydrochloric acid. The excess of ligand was separated by column chromatography through Sephadex® LH-20 (Pharmacia) by elution with methanol; after solvent removal, the fraction containing the tris-chelate complex was purified further using preparative TLC on silicagel (Merck) and a n-butanol-water-acetic acid mixture (8:2:1 v/v/v) as eluent.

Figure 2. Chemical structures of the luminescent $Ru(5\text{-}odap)_3^{2+}$ and $Ru(pzth)_3^{2+}$ oxygen and carbon dioxide indicators, respectively.

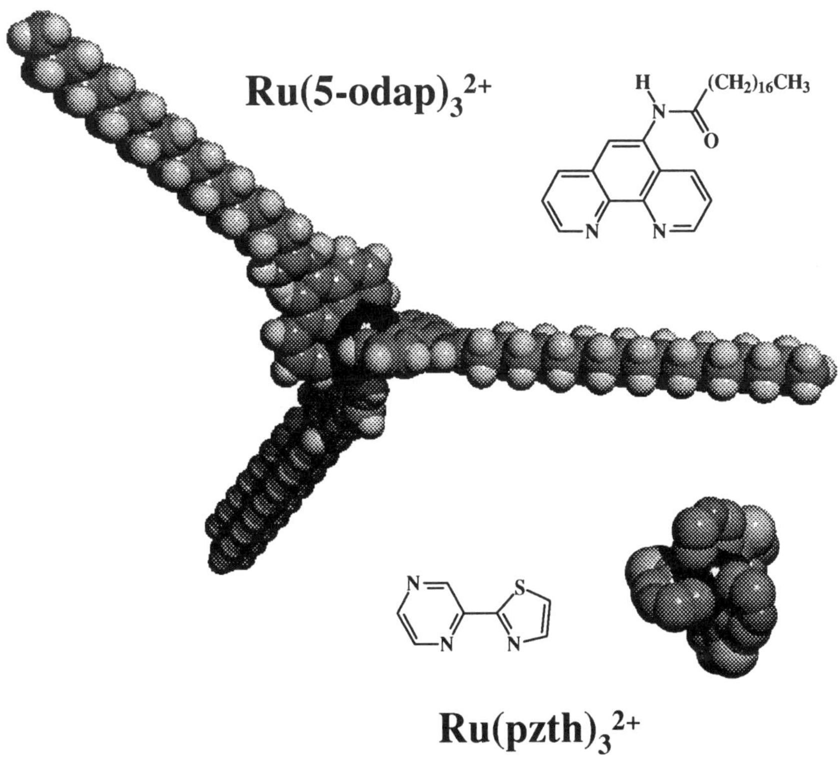

The acetate salt of $Ru(5\text{-}odap)_3^{2+}$ was finally obtained by precipitation of an alcohol solution of the cation with concentrated acetic acid. The purified indicator gave a microanalysis (C, H, N) within 0.4 %. UV-VIS [CH_3OH; λ/nm (ε/L mol^{-1} cm^{-1})]: 449 (17750), 270 (92300), 252 (67700), 220 (76800). The ligand 5-odap was prepared from 5-aminophenanthroline (obtained by chemical reduction of the commercial 5-nitro derivative, Sigma) and stearic acid (Aldrich), using a typical procedure for the synthesis of amides from hindered or unreactive amines via the intermediate carbonate with ethylchloroformate. The synthesis of tris[2-(2-pyrazinyl)-1,3-thiazole]ruthenium(II), Ru(pzth), has been reported already One-component RTV silicone 732 was from Dow Corning. Oxygen, nitrogen and carbon dioxide from cilinders (Air Liquide or Praxair) were mixed within 0.5% of the nominal values using a three-channel mass flow controler electronic unit (ICP, Madrid). The gas mixtures are saturated with water for all the CO_2 measurements and for determinations of dissolved O_2.

For laboratory optimisation of the response, the optical sensing membranes were placed into a flow-through cell previously described, to which the common end of a glass or silica fibre-optic bifurcated bundle was connected. Gas or aqueous samples can be easily measured with such a device.

Two types of portable instruments have been used for the chemical analysis with optodes. The first one is a commercial Guided Wave model 260 spectrometer (El Dorado Hills, CA, USA); this system employs a 5-volt, 20-watt tungsten-halogen lamp as excitation source, and a red-sensitive Hamamatsu photomultiplier tube to detect the emitted light from the remote sensor head, after grating monocromation. A portable Toshiba TX-3200 computer holds the necessary interface card and serves to process and analyse the data. The LAP-1000 is a custom-made laptop fibre-optic spectrometer we have assembled to measure with optodes, both in the laboratory and in the field. Into a 46x33x16 cm aluminum-plastic case, it holds a 12-volt 10-Hz Ocean Optics (Eerbeek, NL) PX-1 xenon flashlamp fitted with an SMA connector for waveguides, and a Oriel (Strattford, Connecticut, USA) Instaspect IV CCD detector (1024 x 256 pixel) attached to a Oriel MS-256 1/8-m spectrograph. The full-length ISA interface card which powers the CCD, fires the strobe, collects the signal and synchronises the excitation and recording events is placed into the mini-docking unit of a Olivetti Echos P-100 laptop computer loaded with the original Oriel software for instrument control and data processing. Both portable spectrometers use the same pair of wide band-pass (400 nm) and cut-on (530 nm) coloured glass filters in the excitation and emission channels, respectively, unless otherwise stated.

4. Results and discussion

Tris(1,10-phenanthroline)ruthenium(II), or $Ru(phen)_3^{2+}$, has been so far one of the most popular luminescent indicators to measure molecular oxygen.[Erreur! Signet non défini.] This analyte is able to quench the fluorescence of the dye with a rate constant of 4 x 10^9 L mol^{-1} s^{-1} in aqueous solution. However, being a water-soluble salt, it is impossible to dissolve it either directly into the hydrophobic silicone membranes (or the like) required to sense oxygen in water, or in a solvent compatible with the organic polymer (halogenated or aromatic hydrocarbons).
Therefore, the luminescent cation needs to be immobilized onto a solid support compatible with the silicone prepolymer, - typically silica gel or glass powder -, previously to fabrication of the membrane. Adsorption of the indicator on the support leads always to inhomogeneous excited-state quenching and strongly non-linear dose–response curves.

Preparation of $Ru(5\text{-}odap)_3^{2+}$ has overcome such problems while maintaining the same interesting indicator features of its phen parent. Figure 3a shows the response function to oxygen in the gas phase of a 150-µm RTV silicone membrane prepared by uptake of $Ru(5\text{-}odap)_3^{2+}$ from a 0.2 mmol L^{-1} solution of the dye in dichloromethane. The inset depicts the associated (linear) calibration curve. Fabrication of a similar membrane with $Ru(phen)_3^{2+}$ using the same procedure is prevented by the almost negligible solubility of this indicator in the required organic solvent. The obtention of a linear dose-response plot allows a two-point calibration procedure to be developed, a highly desirable feature for the commercial aplication of any sensing device. At a total gas flow of 200 cm^3 min^{-1}, the response time for a 90% signal change (t_{90}) is 6 s for a 0–21% O_2 concentration increase and 48 s for a 21–0% reverse variation step. Unlike the oxygen electrode, the optical sensor is able to measure continuously for a long time in the gas phase without damage or drift, since it uses a "dry" sensing membrane and no reference electrode is required.

Figure 3. Response of the Ru(5-odap)$_3^{2+}$-containing luminescence optode to the oxygen concentration in (a) nitrogen and (b) water, at 25 ºC and 711 Torr atmospheric pressure, using the portable CCD spectrometer LAP-1000 (see Experimental Section). Insets: Corresponding dose/response plots (background corrected) in gas.

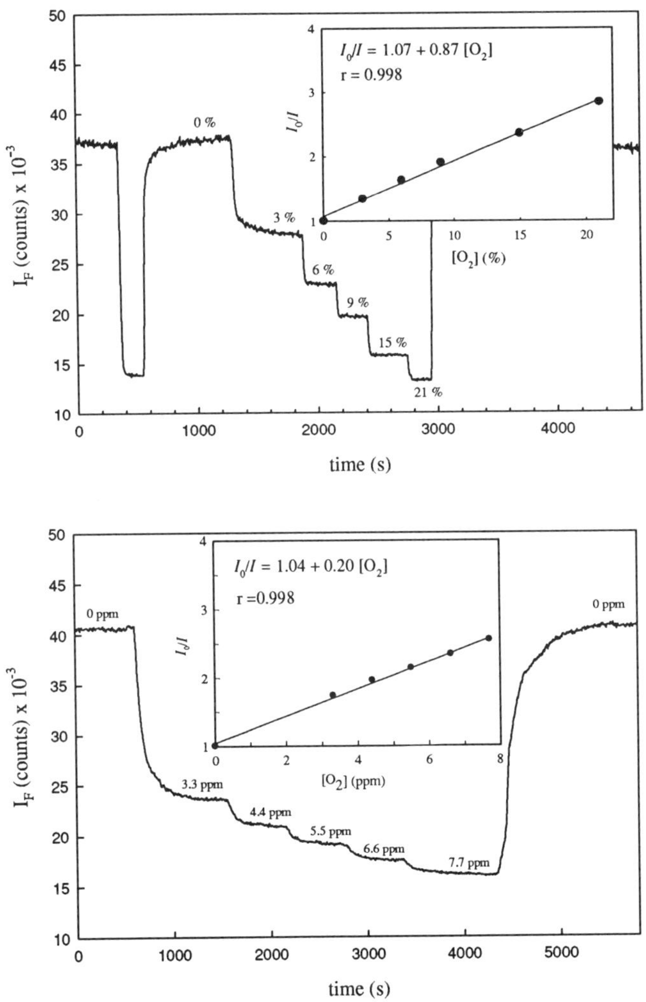

The optode performs equally well in water measurements. Figure 3b shows the sensor response to dissolved oxygen and the corresponding calibration curve. Compared to gas-phase measurements, the longer response time, both to direct and reverse analyte concentration changes (t_{90} of 3.1 and 9.8 min, respectively), is a consecuence of the (slower) diffusion kinetics of molecular oxygen in the aqueous phase and the neccesity of water-saturating thoroughly the gas mixtures before entering the flow-through measuring cell. Moreover, the observed smaller oxygen sensitivity of the optical sensor in water compared to that in gas phase may be a consequence of penetration of water vapour into the indicator layer. The commercial formulation of the silicone prepolymer contains already silica powder as filler, the hydration of which probably reduces the membrane uptake of oxygen, a hydrophobic species.

We have tested one of our oxygen optodes in the in-situ continuous monitorisation of this chemical species into the secondary treatment pool (microorganism-catalysed oxydation) of a municipal water-treatment plant. Figure 4 shows a typical three-day response obtained with the fibre-optic sensor directly introduced into the sewage (Figure 5). For the sake of comparison, hourly measurements were taken by the station employees using a commercial oxygen electrode (Danfoss) permanently installed in the same monitoring site; therefore, no readings have been taking off-hours.

The measurements have been performed with our GW-260 spectrometer (see Experimental Section). In order to correct the readings for variations in the lamp output, mostly due to over 15 ºC temperature difference between day and night, the sensor luminescence signal at 648 nm was always ratioed to the scattered light from the lamp at 520 nm by employing a proper combination of band-pass and cut-on coloured-glass filters (445 and 530 nm, respectively). The data depicted in Figure 4 demonstrates that real time in situ monitoring of dissolved oxygen with optodes is possible indeed; obviously, longer tests and extensive validation have to be performed before a commercial device can be launched, with the essential pressure and temperature corrections incorporated too.

This issues, as well as fabrication of a dedicated portable luminescence spectrometer, are being addressed in our industrial-university R+D project (1997-1999).

Figure 4. Comparison between in-situ measurements of dissolved oxygen in the biological treatment pool of a water sanitation plant taken with the luminescence optode (every 5 min) and a commercial electrode (every 60 min during working hours). The oxygen electrode trace has been offset by –1 ppm for the sake of clarity. The emission intensity in the absence (I_0) and in the presence of the analyte (I) has been corrected for the lamp output (I_r, see text).

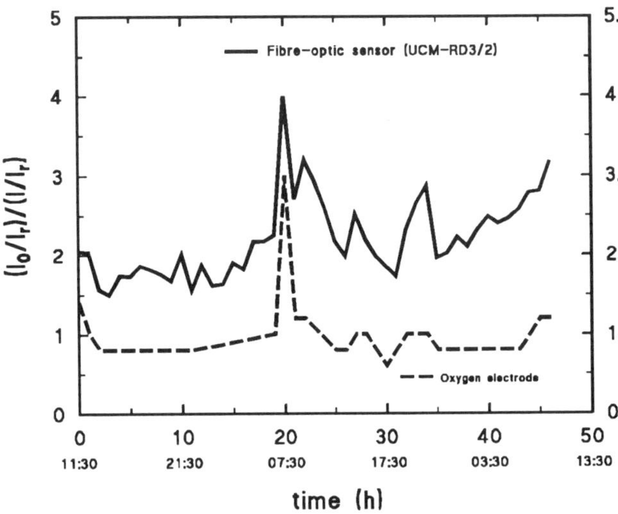

Figure 5. Experimental setup for continuous dissolved oxygen monitoring with optical fibre sensors in the biological treatment pool of a wastewater sanitation plant.

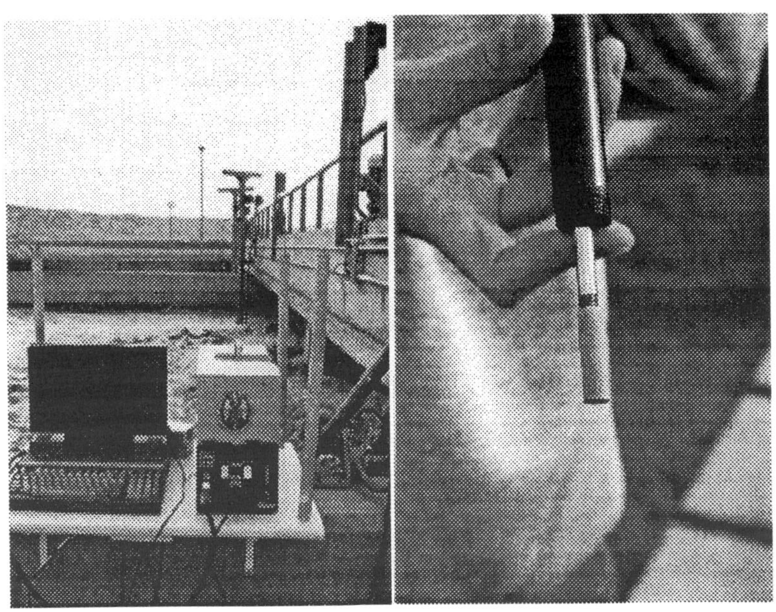

Ru(II) complexes with polyazaheterocyclic ligands can be also engineered to respond to a very different analyte such as carbon dioxide If pyridine-type moieties such as those contained in the 1,10-phenanthroline chelating ligands are replaced by pyrazine rings (Figure 2), the electronically excited state of the corresponding coordination compound turns out sensitive to the solution acidity. Irreversible protonation of the luminescent state competes with radiative deactivation to the ground state, so that both the emission intensity and lifetime are quenched as the pH of the medium is lowered. Permeation of CO_2 through a hydrophobic membrane into an aqueous gel layer containing immobilized $Ru(pzth)_3^{2+}$, placed at the distal end of a fibre-optic bundle, leads to an acidity change thanks to the (reversible) hydration of the analyte gas to hydrogencarbonate ion (Figure 6).

Figure 6. Response of the $Ru(pzth)_3^{2+}$-containing luminescence optode to the carbon dioxide concentration in water-saturated argon, at 25 ºC and 711 Torr atmospheric pressure, using the portable GW-260 spectrometer (see Experimental Section). Insets: Working principle of the optical sensor and corresponding dose/response curve.

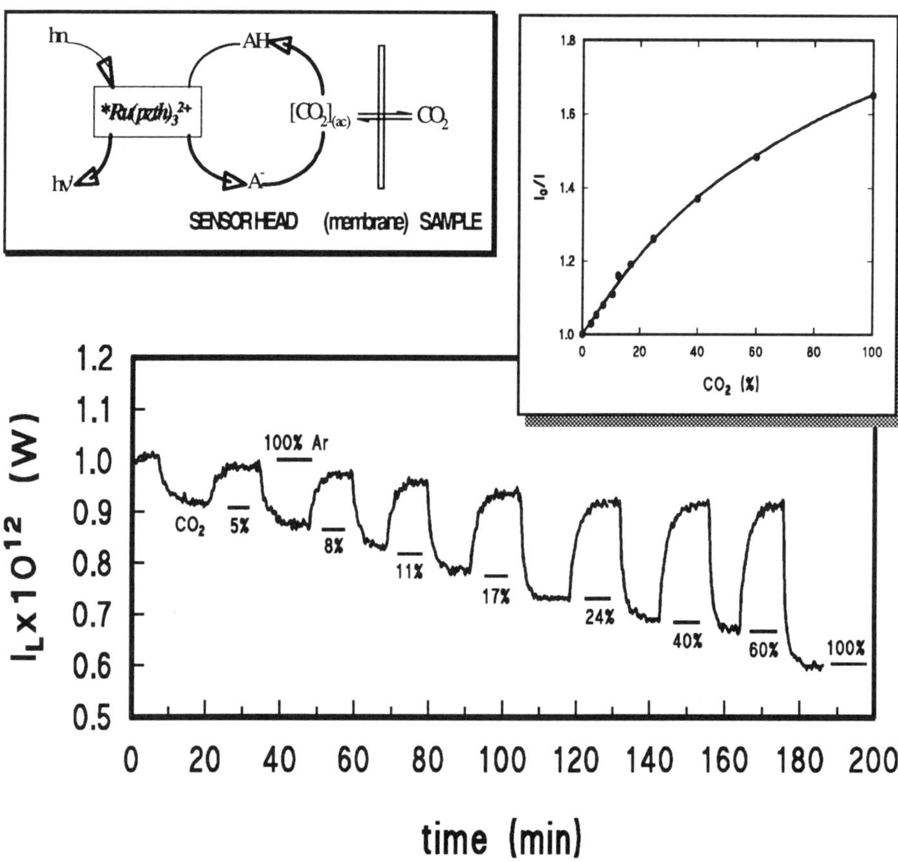

The pH variation is sensed immediately by the luminescent indicator the signal of which, after appropriate calibration, yields the concentration of CO_2 in a gas or liquid phase, or HCO_3^- /CO_3^{2-} dissolved in water upon conversion into CO_2 in a similar (but optical) way to the conventional carbon dioxide (pH) electrode.

Figure 6 depicts the response of the CO_2 optode to the concentration of this determinand in gas mixtures. The response is fully reversible provided the sensor head is kept properly hydrated. This is not a problem for CO_2 monitoring in aqueous samples, but may be a drawback for vapor-phase determinations. The non-linearity of the calibration curve, compared to the response of the oxygen optode, is intrinsic to the acid-base photochemistry involved and not to inhomogeneous environment around the immobilised Ru(II) indicator.
In conclusion, success of fibre-optic sensors for environmental and industrial monitoring depends heavily in the photochemistry involved. Tailor-made optical indicators are required in order to exploit fully the performance of dedicated optoelectronic devices and to jump the gap between academic research and comercial exploitation.

Acknowledgements. Part of this research has been funded by the Spanish Government Agency CICYT (project no. AMB95-0689-C02). D.G.-F. and M.D.M. gratefully thank Universidad Complutense de Madrid for doctoral grants. J.D. acknowledges CICYT for a doctoral grant at INTERLAB Group R+D unit. The authors thank V. Llorente and A. Parrilla of Canal de Isabel II water company for their help and assistance to perform in situ measurements.

REFERENCES

1 Official Diary of the European Communities, no. L 126/1 (18.05.94).
2 O. S. Wolfbeis (editor), "Fiber Optic Chemical Sensors and Biosensors" (vols. 1 and 2), CRC Press, Boca Raton, Florida (USA), 1991.
3 C. Cámara, M. C. Moreno and G. Orellana, "Chemical Sensing with Fiberoptic Devices", in "Biosensors with Fiberoptics", D. L. Wise and L. B. Wingard, Jr. (editors), The Humana Press, Clifton, New Jersey, pp. 29-84, 1991.
4 (a) M. C. Moreno, A. Martínez, P. Millán and C. Cámara, "Study of a pH sensitive optical fibre sensor based on the use of cresol red", J. Molec. Struct. 1986, 143, 553-556 ; (b) M. C. Moreno, M. Jiménez, C. Pérez Conde and C. Cámara, "Analytical performance of an optical pH sensor for acid-base titration", Anal. Chim. Acta, 230, 35-40, 1990.
5 A. Martínez, M. C. Moreno and C. Cámara, "Sulfide Determination by N,N-Dimethyl-p-phenylenediamine Immobilization in Cationic Exchange Resin Using an Optical Fiber System", Anal. Chem., 58, 1877-1881, 1986.
6 (a) M. Faraldos, M. C. Moreno and C. Cámara, "Reversible optical fibre sensor to determine iron", Analusis 1988, 16, 87-91 ; (b) J. M. Barrero, M. C. Moreno-Bondi, M. C. Pérez-Conde and C. Cámara, "A biosensor for ferric ion", Talanta, 40, 1619-1623, 1993.
7 (a) G. Orellana and M. C. Moreno Bondi, "Sensor óptico" ("Optical Sensor"), Span. Patent 9002740 (1990) ; (b) G. Orellana, M. C. Moreno-Bondi, E. Segovia and M. D. Marazuela, "Fiber-Optic Sensing of Carbon Dioxide Based on Excited-State Proton

Transfer to a Luminescent Ruthenium(II) Complex", Anal. Chem, 64, 2210-2215, 1992. ; (c) M. D. Marazuela, M. C. Moreno-Bondi and G. Orellana, "Enhanced performance of a fiber-optic luminescent CO2 sensor using carbonic anhydrase", Sensors & Actuators, B: Chem., 29, 126-131, 1995. (d) G. Orellana, C. de Dios, M.C. Moreno-Bondi and M.D. Marazuela, "Intensity- and Lifetime-Based Luminescence Optosensing of Carbon Dioxide", in "Chemical, Biochemical, and Environmental Fiber Sensors VII", A. M. Scheggi (ed.), Proc. SPIE Vol. 2508, SPIE, Bellingham, WA,; pp. 18-29, 1995.

8 (a) F. J. Mingoarranz, M. C. Moreno-Bondi, D. García-Fresnadillo, C. de Dios and G. Orellana, "Oxygen-Sensitive Layers for Optical Fiber Devices", Mikrochim. Acta, 121, 107-118, 1995. ; (b) G. Orellana, D. García Fresnadillo, M.C. Moreno Bondi and M. P. Xavier, "Método de unión a polímeros de un compuesto indicador sensible al analito, para fabricación de sensores ópticos y su dispositivo" ("Procedure for polymer binding of an analyte-sensitive reagent to fabricate optical sensors and its device"), Span. Pat. (Appl. 9602181, 1996), Universidad Complutense de Madrid, 1996.

9 (a) G. Orellana, M. C. Moreno, C. de Dios, A. A. García and A. M. Gómez, "Sensor óptico luminiscente" ("Luminescent Optical Sensor"), Span. Patent 9300938 (1993) ; (b) G. Orellana, A. M. Gómez-Carneros, C. de Dios, A. A. García-Martínez and M. C. Moreno-Bondi, "Reversible Fiber-Optic Fluorosensing of Lower Alcohols", Anal. Chem., 34, 2231-2238, 1995.

10 G. Orellana, M.C. Moreno Bondi, D. García Fresnadillo, M.D. Marazuela, C. Knapp and A.M. Castro, "Sensor óptico para detergentes" ("Optical sensor for detergents"), Span. Pat. (Appl. 9602182, 1996), Universidad Complutense de Madrid, 1996.

11 C. Cámara, C. Pérez-Conde, M.C. Moreno-Bondi, and C. Rivas, "Fibre-Optical Sensors Applied to Field Measurements", in "Quality Assurance for Environmental Analysis: Method Evaluation within the Measurements and Testing Programme (BCR)", Quevauviller, Ph., Maier, EA. and Griepink, B. (eds.), Elsevier; pp. 165-192, 1995.

12 The optoelectronics partner in the current RTD project is the firm T.G.I. (Technology and Management of Innovation, Velázquez, 134bis, E-28006 Madrid, Spain; c/o Cristina Viguera and José Luis García).

13 W. R. Gruber, I. Klimant and O. S. Wolfbeis, "Instrumentation for optical measurement of dissolved oxygen based on solid state technology", Proc. SPIE, vol. 1885, 448-457, 1993.

14 J.N. Demas and B. DeGraff, "Design and Applications of Highly Luminescent Transition Metal Complexes", Anal. Chem., 63, 829A-831A, 1991.

15 (a) G. Orellana, M.L. Quiroga and A.M. Braun, "Spectroscopic, Electrochemical, and Kinetic Characterization of New Ruthenium(II) Tris-chelates Containing Five-Membered Heterocyclic Moieties", Helv. Chim. Acta, 70, 2073-2086, 1987. ; (b) G. Orellana, C. Alvarez-Ibarra and J. Santoro, "H-1 and C-13 NMR Coordination-Induced Shifts in a Series of Tris(α-diimine)ruthenium(II) Complexes Containing Pyridine, Pyrazine, and Thiazole Moieties", Inorg. Chem., 27, 1025-1030, 1988. ; (c) G. Orellana, C. Alvarez-Ibarra and M.L. Quiroga, "New Binuclear Heterocyclic Ligands Sharing an a-Diimine Moiety and Their Ruthenium(II) Tris-chelates", Bull. Soc. Chim. Belg., 97, 731-741, 1988. ; (d) G. Orellana and A.M. Braun, "Quantum Yields of 3MLCT Excited State Formation and Triplet-triplet Absorption Spectra of Ru(II) Tris-chelate Complexes Containing Five- and Six-membered Heterocyclic Moieties", J. Photochem. Photobiol., A: Photochem., 48, 277-289, 1989. ; (e) M.C. Moreno-Bondi, G. Orellana, N.J. Turro and D.A. Tomalia, "Photoinduced Electron Transfer Reactions to Probe the Structure of Starburst Dendrimers", Macromolecules, 23, 910-912, 1990. ; (f) G. Orellana,

M.L. Quiroga and C. de Dios, "Spectroscopy, Electrochemistry, and Photochemistry of Ruthenium(II) Complexes Containing Five-Membered Heterocyclic Chelating Ligands", Trends Inorg. Chem., 3, 109-130, 1993. ; (g) J.-P. Lecomte, A. Kirsch-De Mesmaeker and G. Orellana, "Photophysics of Polyazaaromatic Ruthenium(II) Complexes Interacting with DNA", J. Phys. Chem., 98, 5382-5388, 1994.

16 D. García-Fresnadillo, "Molecular engineering and photochemistry of Ru(II) luminescent indicators for oxygen, detergents and pesticides optosensing", Ph.D. Thesis, Universidad Complutense de Madrid, Madrid, p. 186, 1996.

17 D. García-Fresnadillo, Y. Georgiadou, G. Orellana, A. M. Braun and E. Oliveros, "Singlet Oxygen (1Δg) Production by Ruthenium(II) Complexes Containing Polyazaheterocyclic Ligands in Methanol and in Water", Helv. Chim. Acta, 79, 1222-1238, 1996.

18 The measurements were performed at "Soto de Gutiérrez" wastewater-treatment station nearby Madrid, operated by Canal de Isabel II water company, which sanitises the efluents from Ciempozuelos, San Martín de la Vega and Valdemoro municipalities.

OPTOCHEMICAL SENSORS IN WATER MONITORING

W. TRETTNAK and F. REININGER, Joanneum Research, Institute for Chemical and Optical Sensors, Steyrergasse 17, A-8010 Graz, Austria

1. Abstract

Optochemical sensors or optodes have shown to be a suitable tool for the detection of a large number of chemical parameters. They may offer advantages over already existing and established measurement methods, and new analytes have become accessable. Despite these facts, only a small number of optochemical sensors is already available on the market or close to the point of commercialization. In this review, a state-of-art and outlook on the developments in optical sensor instrumentation and the problems associated with it will be presented mainly on the basis of the results obtained at Joanneum Research, and with respect to the application of the sensors to water monitoring.
Optochemical sensors for the parameters oxygen, pH, carbon dioxide and ammonia will be discussed with an outlook on the determination of new species. Some of the critical points in sensor development and some needs of research institutions will be identified aswell.

2. Introduction

A considerable increase in research activity in the field of the development of optochemical sensors (or "optodes") could be observed in the last two decades. This fact is supported by an annually and steady increasing number of publications and conferences on this subject. The number of research groups all over the world seems to be augmented, new small companies have been founded, and many large companies are known to have been experimenting or at least "playing" with optochemical sensors. It has been demonstrated that optochemical sensors can measure numerous of parameters and are capable of offering a true alternative over already existing and established methods. For example, in contrast to electrochemical sensors, optochemical sensors do not require an electrolyte solution, do not consume the analyte and are not affected by electromagnetical fields.

However, there is a large discrepancy between the number of optical sensors, which have been reported and investigated in the laboratories, and the number of sensors, which are already available on the market. Only a few optochemical sensors are commercially available, mainly sensors for basic analytes, such as oxygen, pH and carbon dioxide. These sensors have been developed with a special view on medical application in blood gas analysis or blood gas monitoring. Only recently, optochemical sensors have also been offered for water monitoring and environmental analysis. One of the reasons may be that optical sensor development is an interdisciplinary field which needs inputs from, for example, chemistry, optics and electronics.

This requires heterogeneous research groups or well-established cooperation networks, which are both still in an early stage of development.

Joanneum Research, which is Austrias second-largest research institution, has been active in the field of optical sensor development for several years. Optochemical sensors have been investigated and developed for various parameters (e.g. pH, carbon dioxide, ammonia) with the main focus on oxygen measurement and its application to various fields. In this review, a summary will be given on the developments and experiences obtained in optochemical sensors on the background of our results and with respect to their application in water monitoring. The aim is to provide a better understanding in sensor development, to identify critical points and the needs for future sensor development coming from the research institutions, the industry (which should provide the boundary conditions) and, finally, the end user.

3. Optochemical sensors for oxygen

Most optochemical oxygen sensors are based on the quenching of the luminescence of an oxygen-sensitive dye by molecular oxygen. A variety of dyes and measurement schemes have been reported to date .
Typically, the dye is incorporated into a gas-permeable membrane, and the luminescence intensity or the luminescence decay time, which are both dependent on the oxygen concentration, are determined [1].
The relationship between the oxygen concentration and the luminescence intensity or luminescence decay time can (in an ideal case) be described by the Stern-Volmer equation [1] :

$$\frac{\tau_0}{\tau} = \frac{I_0}{I} = 1 + K_{sv} \cdot [O_2] \qquad \text{Eq. 1}$$

where I_0 and I are the luminescence intensities and τ_0 and τ are the luminescence decay times in absence and in presence of oxygen at a concentration $[O_2]$, respectively. K_{SV} is the quenching or Stern-Volmer constant and is a measure for the oxygen sensitivity of the dye. Both I_o or τ_0 and K_{sv} are dependent on temperature. The intensity ratio I_0/I or the decay time ratio τ_0/τ can directly be related to the oxygen concentration.

The measurement of luminescence intensity is usually performed with bulky instrumentation such as expensive fluorescence spectrometers or fibre-optic photometers, which are of limited applicability. Recently, oxygen measurement systems have been presented, which are based on low-cost semiconductor devices, such as light-emitting diodes (LEDs), photodiodes and low-cost analogue and digital components. Joanneum Research has developed such an oxygen sensor instrumentation, which is suitable for the determination of oxygen in gases or dissolved in liquids [2]. A ruthenium(II) complex, which is immobilized in a polystyrene layer, is used as the oxygen-sensitive dye.
A blue light-emitting diode is used for excitation of the luminescence, whereas a photodiode serves as the detector. The cell geometry has been optimized for signal yield, while minimizing straylight. The flow-through cell (the volume is approximately 50 µl) is compact, robust and

can be thermostatted by heating. It has been designed for sample flow rates of 0.1 to 2 ml/min. The measurement instrument is equipped with a built-in pressure sensor and an additional external temperature sensor, which are both needed for calibration of the device. Sensor calibration is performed with two solutions, an oxygen-free sulfite solution and air-saturated water. This two-point calibration proved to be sufficient for most of the water-monitoring applications.

The instrument has been applied to monitoring of oxygen in a lysimeter station, where ground water is drawn continuously from the soil at certain depths during its infiltration process. The oxygen concentration in the infiltration water can be considered as an indirect measure for the microbial and biochemical activity in the soil. The amount of water, which is collected, is small and may vary considerably with weather and season. This makes it difficult to use conventional polarographic oxygen sensors to determine the oxygen concentration. The optochemical oxygen sensor, which does not consume any oxgyen, is not affected by varying sample flow rate or drying out of the sensor membrane. Furthermore, this type of sensor is less affected by gases such as carbon dioxide or hydrogen sulphide and by membrane fouling. The measurement setup used in the lysimeter station is shown in Figure 1a. (see Figure 1a,b in annex)

Initially, the optical oxygen sensor was calibrated weekly. However, thermostatization by heating of the flow-through cell to a few degrees Celsius above ambient temperature was not convenient, since degassing of the sample solution lead to erroneous results. For this reason, the cell thermostatization was switched off and all calibrations were performed at ambient temperature. Since the temperature characteristics of the sensor were unknown, the temperature dependencies of the calibration parameters I_0 and K_{sv} were calculated after a measurement period of approximately half a year from the data, which were collected at the weekly calibrations. This was possible under the assumption that the sensor characteristics remained unchanged over this time period and that no photobleaching of the dye occurred. This was also veryfied for measurement periods of one to two months with calibrations at constant temperature. A comparison of the results obtained within the measurement period of March to August 1994 and of the oxygen concentrations of air-saturated soil infiltration water, which were calculated on the basis of the water temperature and of the barometric pressure, are shown in Figure 1b. The oxygen concentration was found to be always above 5 mg/l (i.e. more than 50 % air satuartion), although it decreased continuously within the observation period. This may be related to oxygen-consuming processes, which may be attributed to increasing biological and biochemical activity in the soil. Infiltration of rain water with higher oxygen content was identified as the main reason for short term fluctuations [3]. (see Figure 2 in annexe)

The measurement of luminescence intensity may be affected by e.g. changes in the light source intensity, by photobleaching or leaching of the dye, by intrinsic sample fluorescence or other optical properties of the sample solution. Many of these problems can be overcome by measuring the luminescence decay time of the oxygen-sensitive dye. Systems based on the measurement of the luminescence decay time offer the advantage that they are (in the first order) independent from the light intensity. This is of importance, since it reduces many requirements concerning the optical components, e.g. it becomes feasible to measure via optical fibres, where the optical attenuation may be unknown or even variable. Furthermore, it becomes possible to measure in closed systems through the walls of a transparent vessel ("contactless sensing").

The decay time of the luminescence can be determined directly by employing a pulsed light source and high-speed detectors (with a sampling frequency of approximately 10 MHz). Such equipment, which is typically based on a laser, a photomultiplier and an oscilloscope, is large, expensive and cannot be considered practical for many purposes [4]. Another approach to the determination of luminescence lifetimes is based on phase modulation techniques [5-8]. In the meantime, several instruments of phase-modulated phosphorimeters and fluorimeters have been described. Oxygen measurement systems based on the use of an LED as the light source and a photomultiplier as the detector have been reported for oxygen-sensitive ruthenium-complexes, which have fluorescence lifetimes of up to approximately 5 μs [9 - 11]. In case that luminescent dyes with lifetimes in the range of tens and hundreds of microseconds (e.g. porphyrin dyes) are used, the required modulation frequencies can be essentially lowered. Thus, a substantial simplification of the system electronics becomes possible [12 - 14].

A cheap and simple oxygen measurement system has been presented, where the phase measurement is performed at a single frequency only and by measuring the delay time between a square-wave excitation and a square-wave in the detector, generated via suitable amplification of the emission signal coming from the sensor [15]. The instrument is based on low-cost components such as LEDs and photodiodes with the measurement module having an overall size of only 120x60x30 mm. Platinum(II) octaethylporphyrin-ketone is used as the oxygen-sensitive dye. It has an unquenched lifetime of approximately 60 μs and can be excited with a yellow/amber LED as the light source [4,13]. The instrument has been designed for either measurement of oxygen in gases or (in combination with a flow-through-cell) in liquids.

With this instrument, dissolved oxygen could be determined in concentrations of up to air saturation (i.e. approximately 0 to 200 hPa oxygen partial pressure) with an accuracy of better than $\pm$ 1 hPa for measurements over periods of typically 24 hours with the whole device being thermostatted (Figure 2). Due to the high oxygen sensitivity of the dye (when immobilized in a high-molecular weight polystyrene matrix), the resolution in oxygen reading was improved when compared to the previously used ruthenium(II) complex [2]. A resolution of 0.1 hPa was obtained at low oxygen concentrations. The small dimensions of the primary measurement unit (120x60x30 mm) are a substantial improvement, when compared to previously reported optical oxygen sensor units [2]. The new measurement technique, which is based on a luminescence decay time measurement, has proved to be clearly superior to intensity-based methods. (see Figure 2 in annexe)

Based on these results and new developments in measurement electronics, a new oxygen sensor device is now in development, which is designed for monitoring of oxygen in waste water, groundwater (e.g. in bore holes) and surface water. The optode has the same shape and dimensions as the commercially available oxygen electrodes CellOx 700 or TriOxmatic 700 from WTW (Weilheim, Germany) (Figure 3). The optical cell and a signal pre-amplification are integrated in the head of the probe, which is connected to a data processing unit by means of a cable. The final aim is to obtain an oxygen sensor, which is capable of measuring under the conditions, which are specified in Table 1, and can be regarded as the ideal and the minimum requirements for environmental application of an oxygen sensor. (see Figure 3 and Table 1 in annexe)

From these specifications, it can be seen that the requirements are quite tough in order to be able to compete with already existing electrode systems. A high measurement stability is the pre-requisite for environmental application. Factory pre-calibration is desired and a single-point calibration (e.g. under air), as it is usually performed with oxygen electrodes, may be acceptable for the customer.

The main advantage of the optochemical oyxgen sensor (when compared to amperometric sensors) is its insensitivity to substances such as hydrogen sulfide, which have normally to be encountered in waste water monitoring. The absence of electrolyte solution will simplify the exchange of sensor membrane and the sensor handling. Furthermore, there is no long stabilization time necessary for the optical sensor, whereas oxygen electrodes may need up to hours to give a stable signal after switching on the instrument (the so-called "polarization time").

Despite the requirements on optical, electronical and mechanical design, it has to be pointed out that the major critical point is probably the oxygen sensor membrane, which has to be characterized very well. The oxygen response characteristics of this membrane, which have to be known in the whole temperature range of interest in order to allow for a temperature compensation, should not change over up to one year of continuous use under varying environmental conditions. Furthermore, they also should not change over a certain period of storage time before being used. This shelf lifetime should also be in the order of years.

Whereas the above-described oxygen sensor may be considered as a "macrooptode", oxygen microoptodes have been developed at the Max-Planck-Institute for Marine Microbiology in Bremen (Germany) and at the University of Regensburg (Germany) [16,17]. The microoptodes (Figure 4a) can be used for measuring oxygen gradients e.g. in marine and freshwater sediments with a high spatial resolution (better than 50 μm). The sensors are based on tapered silica/silica multimode fibers with tip diameters of 5 to 100 μm (depending on the type of application). Depending on the indicator and the polymer used, micropotodes with different measurement ranges can be obtained. A comparison of oxygen microelectrodes and oxygen microoptodes shows that the properties of the optical sensor are superior to that of the electrodes (Table 2). Whereas luminescence intensity measurement has been used at the beginning [16], the new generation of microoptodes is also based on a phase measurement technique [17]. A typical measurement of an oxygen profile in a sediment with both an oxygen microoptode and an oxygen microelectrode is shown in Figure 4b. (see Figure 4a,b and Table 2 in annexe)

4. Optochemical sensors for pH, ammonia and carbon dioxide

Most optochemical sensors for the parameters pH, ammonia and carbon dioxide have in common that a pH-sensitive dye is used as the indicator [1]. Typically, the colour or the luminescence properties of the dye are changed in presence of the analyte, which can be quantified e.g. by measuring the absorbance or luminescence intensity. In the case of optical pH sensors, the indicator dye is immobilized in a hydrophilic polymer or on the surface of a suitable substrate. A large variety of pH-sensitive dyes has been used to produce pH optodes [1]. In contrast to pH electrodes, no electrolyte solution is required with the optical sensor, no reference electrode is needed and the sensor is electrically isolated from the sample solution. Although pH electrodes are well-established and highly-developed, still new pH sensors for environmental application are of interest, which show a high measurement stability and which can be used at low sample conductivity (e.g. in rain water or most of the spring waters).
Optical pH sensors have been developed at Joanneum Research, where a pH-sensitive dye is immobilized onto a water-permeable celluloseacetate support [18,19]. The resulting sensor membranes change their colour from yellow to blue in a range of approximately pH 5 to 10 (Figure 5a) and show extremely good storage and measurement stability. A resolution of pH 0.001 could be obtained at the pK of the dye by measuring the transmittance of the membranes. Accuracies of pH $\pm$ 0.04 and of $\pm$ 0.08 were obtained immediately after and within one month after a four-point calibration, respectively.

The response times, which are dependent on the buffer capacity of the sample solution, were less than one minute for 99 % of the signal change in 0.1 M buffer solutions. Despite these very promising sensor properties, it was found that the sensor response curve is strongly affected by the ionic strength of the sample solution (Figure 5b). For application of the sensor, this would mean that the sample ionic strength or its range of variation should be known, and that a constant or a (more or less constant) high ionic strength would be required for pH measurement. This is the case, for example, in sea water. Since the pK of any pH-sensitive dye (whether dissolved or immobilized on a solid matrix) is influenced by changes in the ionic strength of the sample solution, the development of optochemical pH sensors for environmental application with varying or unknown composition of the sample solution (e.g. waste water) turns out to be quite difficult. The problem may be minimized by choosing a dye with low sensitivity to ionic strength in combination with a matrix, which additionally reduces these ionic strength effects. Furthermore, it has to be considered that, for practical purposes, the optical pH measurement has to be optically isolated from the sample solution, since it may be influenced by sample colour, turbidity or intrinsic sample luminescence. (see Figure 5a,b in annexe)

In the more "classical" approach for ammonia and carbon dioxide measurement, the pH indicator is separated from the sample solution by means of a gas-permeable membrane [1]. Ammonia and carbon dioxide diffuse through this membrane into an internal buffer solution, where the pH is changed. This again causes a change in the spectral properties of the indicator. The principle can be compared to the potentiometric ammonia and carbon dioxide electrodes, where a pH electrode is inserted into an internal buffer solution. For this reason, optical sensors based on this measurement principle are also prone to similar influences than the corresponding electrodes and show similar disadvantages. A change in sample ionic strength leads to a difference in osmotic pressure between sample solution and internal buffer solution. This causes diffusion of water into or out of this internal buffer solution, which again results in drifts in the sensor signal and changes in the sensitivity for the analyte. However, optical sensors of this type have shown to be suitable for applications, where the ionic strength of the sample solution remains rather constant [20]. This is the case especially in sea water monitoring, where a high sensitivity to CO_2 is required, which cannot be obtained with conventional CO_2 electrodes. Optical carbon dioxide sensors have been developed by various institutions especially for this application and tested successfully [21-24].

In the past years, direct immobilization of the pH-sensitive dyes in a gas-permeable polymer matrix has shown to be a suitable technology for the preparation of carbon dioxide [25-27] and ammonia sensors [28,29]. This is achieved in most cases by the formation of ion pairs between the dye and lipophilic counter ions (e.g. quaternized alkyl ammonium salts). The sensor membranes obtained in this way do not require an internal buffer solution and are not influenced by varying sample osmolarity or ionic strength. This simplifies the sensor production substantially.
The sensors are based on the measurement of absorbance, luminescence intensity and even luminescence decay time, although the decay times utilized are still too short (i.e. in the range of a few nanoseconds) for realization of a simple and cheap sensor instrumentation [30,31].

Joanneum Research has been following this new approach with the aim of the application of the carbon dioxide and ammonia sensors to water monitoring [32-34]. Figure 6 shows a typical carbon dioxide measurement performed with a sensor, which was based on the fluorescent dye 8-hydroxypyrene-1,3,6-trisulfonate (HPTS) immobilized in ethylcellulose. The sensor membrane was protected from the sample solution by an additional chemically and mechanically resistant gas-permeable membrane. The fluorescence measurement was performed with a fibre-optic setup in combination with a flow-through cell, which was used for testing purposes.

The sensor provided a linear calibration graph and measurement stability over short periods of up to a few days. At present, this type of CO_2 sensor is in investigation with respect to key features such as the measurement and storage stabilities and their improvement. The ammonia sensor is based on the dye bromothymol blue, which is immobilized (again as an ion pair) in a thin layer of silicone, and the change in colour from yellow to blue in presence of ammonia is monitored. Again the sensor was basically characterized in the laboratory and very promising results were obtained. However, for practical application in environmental sensing (especially for long-term monitoring), the sensor still needs improvement with respect to its measurement stability, which is limited to a few days only [34]. (see Figure 6a,b in annexe)

Both, the carbon dioxide and ammonia sensors may provide several advantages over the commercially available electrodes, which are laboratory devices only, and which are not intended to be used in field applications. The optical sensor membranes can be prepared very easily and on a cheap cost basis. Corresponding low cost instrumentation for this type of sensors is also in development. The instrument prototype will be based on solid state components in combination with fibre optics. Such instrumentation can also be easily used with sensor membranes for other analytes (e.g. ionic species such as heavy metal ions) by adapting the optical components (LEDs, photodiodes, filters) to the spectral characteristics of the respective sensor membrane. Again, the availability of sensor membranes with long luminescence decay times (in the range of μs to ms) as the measurement parameter would allow to simplify the optical setup considerably.

5. Conclusions and outlook

Optochemical sensors have shown to be a suitable tool in monitoring of analytical parameters relevant for the determination of water quality. For specific applications, they may offer advantages over already existing measurement methods. Besides the sensors presented in this review, optochemical sensors have also been reported for numerous other parameters, for example, for gases (e.g. SO_2 and Cl_2), ions (e.g. NH_4^+, Na^+, K^+, Ca^{2+} and Cl^-), heavy metal ions (e.g. Ag^+, Pb^{2+} and Hg^{2+}) and organic substances, such as hydrocarbons, glucose, lactate, amino acids, biodegradable material in general etc..
New technologies and materials are becoming available every year: new dyes and polymers, light sources (e.g. LEDs and laser diodes), integrated optics and microsystems. This is also leading to continuous improvements in already existing sensor systems, and new sensors are becoming available for new parameters and for new applications.

What is required now, is a clear identification of the needs in water monitoring. It has to be identified, which parameters are of analytical interest and for which parameters already existing methods may be regarded as insufficient. It should be clearly stated, what the requirements (depending on the type of application) on these new sensors are, what sensor characteristics they should have in order to be useful for practical purposes, and which sensor characteristics of already existing sensors should be improved. These needs, which should be identified by scientists, the companies, which are already producing sensors, national and international institutions and the end users, should be clearly summarized in a kind of catalog or "hot list". This catalog would enable researchers immediately to focus on sensors for the right parameters, and to avoid developments, which may be considered as a "cul-de-sac", because they are of no practical use.

In a second step, the various research groups (i.e. at the universities, at private and (inter)national research institutions and in the companies) dealing with sensor systems and working on methods for the determination of specific parameters should be identified and also listed in a second part of the catalog. Furthermore, companies, which are able to produce and market sensors as well as the potential end users should also be included. This could lead to the formation of networks and clusters covering basic research, production facilities and the marketing of the final product. Of course, it would be necessary to have annual revision of the needs, the active groups in the field, the companies and the state-of-art. This would, finally, also allow to revise the system of financement of research and sensor development and to help to identify the critical steps from the point of view of the costs. For this purpose, committees

could be formed within the networks or clusters, which are responsible for coordinating or reviewing specific tasks or sections in sensor development.
The universities and institutions, which do the basic research, are mainly interested in developing and demonstrating new sensor principles, and very often stop their work with the publication of some results. The work is usually stopped at a point, where further investigations become tiresome and sometimes, from a "scientists point of view", also boring. To identify sensor membranes with properties, which make them suitable for practical application, to characterize these membranes and to improve these characteristics step by step means just a hell lot of work. The costs for this work are very often too high to be carried by small companies, and the large companies have the tendency to wait until a sensor system is really functioning. Once a working prototype system is available, it is quite easy to find some interest from the industrial side in commercializing a system. However, there is a big gap between the basic results and a working prototype measurement system, which very often can only be closed by contributions from national or international fundings, although contributions from industrial side would be highly desirable. For this reason, it is essential to avoid that several groups are working on the same subject (if there is no real need for it), or that identical work is done in parallel.

Cost effectiveness is maybe one of the prerequisites for future sensor development, especially with respect to the steadily increasing costs, the decreasing amounts of money available for research activities and to the large future potential, which is available, for example, in the field of optochemical sensing.

6. Acknowledgements

The oxygen measurements at the lysimeter station in Wagna (Leibnitz, Austria) were performed in collaboration with the Institute of Hydrogeology and Geothermics (Joanneum Research, Graz, Austria). The new oxygen optode for environmental application has been designed in collaboration with WTW (Weilheim, Germany).

REFERENCES

[1] O. S. Wolfbeis (ed.), Fiber optic chemical sensors and biosensors, Vols. I & II, CRC Press, Boca Raton, FL, 1991.

[2] W. Trettnak, W. Gruber, F. Reininger and I. Klimant, Recent progress in optical oxygen sensor instrumentation, Sensors and Actuators B, 29 (1995) 219-225.

[3] J. Fank, W. Gruber, F. Reininger and W. Trettnak, Die Messung von O_2 im Sickerwasser als Parameter zur Erfassung von bio- und hydrochemischen Umwandlungsprozessen, 5. Gumpensteiner Lysimetertagung, BAL Gumpenstein (Austria), 25.-26. April 1995, 141-145.

[4] P. Hartmann and W. Trettnak, Effects of polymer matrices on calibration functions of luminescent oxygen sensors based on porphyrin-ketone complexes, Anal. Chem., 68 (1996), 2615-2620.

[5] Z. Gaviola, Ein Fluorometer. Apparat zur Messung von Fluoreszenzabklingungszeiten, Z. Physik, 42 (1926), 853-861.

[6] F. Duschinsky, Eine allgemeine Theorie der zur Messung sehr kurzer Leuchtdauern dienenden Versuchsanordnungen (Fluorometer), Z. Physik, 81 (1933), 23-42.

[7] E. A. Bailey and G. K. Rollefson, The determination of fluorescence lifetimes of dissolved substances by a phase shift method, J. Chem. Phys., 21 (1953), 1315-1326.

[8] A. Schmillen, Abklingzeitmessungen an flüssigen und festen Lösungen mit einem neuen Fluorometer, Z. Physik, 135 (1953), 294-308.

[9] M. E. Lippitsch, J. Pusterhofer, M. J. P. Leiner and O. S. Wolfbeis, Fibre-optic oxygen sensor with the fluorescence decay time as the information carrier, Anal. Chim. Acta, 205 (1988) 1-6.

[10] S .B. Bambot, R. Holavanahali, J. R. Lakowicz, G. M. Carter and G. Rao, Phase fluorometric sterilizable optical oxygen sensor, Biotechnology and Bioengineering, 43 (1994) 1139-1145.

[11] G. A. Holst, T. Köster, E. Voges and D. W. Lübbers, FLOX - an oxygen-flux-measuring system using a phase-modulation method to evaluate the oxygen-dependent fluorescence lifetime, Sensors and Actuators B, 29 (1995) 231-239.

[12] D. B. Papkovsky, New oxygen sensors and their application to biosensing, Sensors and Actuators B, 29 (1995) 213-218.

[13] D. B. Papkovsky, G. V. Ponomarev, W. Trettnak and P. O'Leary, Phosphorescent complexes of porphyrin ketones: optical properties and application to oxygen sensing", Anal. Chem., 67 (1995), 4112-4117.

[14] W. R. Gruber, P. O'Leary and O.S. Wolfbeis, Detection of fluorescence lifetime based on solid state technology and its application to optical oxygen sensing, Proc. SPIE, Vol. 2388 (1995) 148-158.

[15] W. Trettnak, C. Kolle, F. Reininger, C. Dolezal and P. O'Leary, Miniaturized luminescence lifetime-based oxygen sensor instrumentation utilizing a phase modulation technique, Sensors and Actuators B, 35-36 (1996) 506-512.

[16] I. Klimant, V. Meyer and M. Kühl, Fiber-optic oxygen microsensors, a new tool in aquatic biology, Limnol. Oceanogr., 40(6) (1995) 1159-1165.

[17] Oxygen microoptodes - the new alternative, (1997) Precision Sensing GmbH, Neuburg a.d.D, Germany.

[18] A. Holobar, R. Benes, B. H. Weigl, P. O'Leary, P. Raspor and O. S. Wolfbeis, Fiber optic and non-fiber optic double beam pH sensor for use in a flow-through cell, Anal. Methods and Instrumentation, Vol. 2, No. 2 (1995) 92-100.

[19] A. Holobar, B. H. Weigl, W. Trettnak, R. Beneš, H. Lehmann, N. V. Rodriguez, A. Wollenschlager, P. O'Leary, P. Raspor and O. S. Wolfbeis, Experimental results on an optical pH measurement system for bioreactors, Sensors and Actuators B, 11 (1993) 425-430.

[20] B. H. Weigl, A. Holobar, N. V. Rodriguez and O. S. Wolfbeis, Chemically and mechanically resistant carbon dioxide optrode based on a covalently immobilized pH indicator, Anal. Chim. Acta, 282 (1993) 335-343.

[21] D. R. Walt, G. Gabor and C. Goyet, Multiple-indicator fiber-optic sensor for high-resolution pCO_2 sea water measurements, Anal. Chim. Acta, 272 (1993) 47-52.

[22] C. Goyet, D. R. Walt and P. G. Brewer, Development of a fiber optic sensor for measurement of pCO_2 in sea water: design and sea trials, Deap-sea research A, 39 (1992) 1015-1026.

[23] M. D. Grandpre, T. R. Hammar, S. P. Smith and F. L. Sayles, In situ measurements of sea water pCO_2, Limnol. Oceanogr., 40(5) (1995) 969-975.
[24] M. D. Grandpre, Measurement of seawater pCO_2 using a renewable-reagent fiber optic sensor with colorimetric detection, Anal. Chem., 65 (1993) 331-337.
[25] A. Mills, Q. Chang and N. McMurray, Equilibrium studies on colorimetric plastic film sensors for carbon dioxide, Anal. Chem., 64 (1992) 1383-1389.
[26] A. Mills and Q. Chang, Fluorescence plastic thin-film sensor for carbon dioxide, Analyst, 118 (1993) 839-843.
[27] A. Mills and Q. Chang, Colorimetric polymer film sensors for dissolved carbon dioxide, Sensors and Actuators B, 21 (1994) 83-89.
[28] W. Sellien, R. Czolk, J. Reichert and H. J. Ache, Development of an optical-chemical sensor for the detection of ammonium ions, Anal. Chim. Acta, 269 (1992) 83-88.
[29] T. Werner, I. Klimant and O. S. Wolfbeis, Ammonia-sensitive polymer matrix employing immobilized indicator ion pairs, Analyst, 120 (1995) 1627-1631.
[30] J. Sipior, S. Bambot, M. Romauld, G. M. Carter, J. R. Lakowicz and G. Rao, A lifetime-based optical CO_2 gas sensor with blue or red excitation and stokes or anti-stokes detection, Anal. Biochem., 227 (1995) 309-318.
[31] Q. Chang, J. Sipior, J. R. Lakowicz and G. Rao, A lifetime-based fluorescence resonance energy transfer sensor for ammonia, Anal. Biochem., 232 (1995) 92-97.
[32] B. H. Weigl and O. S. Wolfbeis, New hydrophobic materials for optical carbon dioxide sensors based on ion pairing, Anal. Chim. Acta, 302 (1995) 249-254.
[33] B. H. Weigl, A. Holobar, W. Trettnak, I. Klimant, H. Kraus, P. O'Leary and O. S. Wolfbeis, Optical triple sensor for measuring pH, oxygen and carbon dioxide, J. Biotecholology, 32 (1994) 127-138.
[34] M. Trinkel, W. Trettnak, F. Reininger, R. Beneš, P. O'Leary and O. S. Wolfbeis, Study of the performance of an optochemical sensor for ammonia, Anal. Chim. Acta, 320 (1996) 235-243.

ANNEXES

Table 1. Main requirements on the performance of an optical oxygen sensor for environmental monitoring.

Dimension	ideal	minimum requirement
dissolved oxygen measuring range	0 .. 1000 hPa pO_2 or 0 .. 50 mg/l O_2	0 .. 500 hPa pO_2 or 0 .. 25 mg/l O_2
measuring accuracy	0.2 hPa pO_2	2 hPa pO_2
calibration	factory pre-calibration	factory pre-calibration + 1-point calibration by user
drift with air saturation	< 1 %/month	< 2 %/month
service life		1 year
response time (system)	t_{90} < 10 s (20 °C) t_{99} < 60 s (20 °C)	t_{90} < 30 s (20 °C) t_{99} < 180 s (20 °C)
sample flow rate		0 ... 5 m/s

Dimension	ideal	minimum requirement
measuring temperature	0 .. 80°C	5 .. 40°C
ambient pressure	10 bar	5 bar
no sensitivity to:	CO_2 (5 g/l) H_2S (500 mg/l) NH_3 (100 mg/l)	CO_2 (2 g/l) H_2S (100 mg/l) NH_3 (50 mg/l)

Table 2. Comparison of different types of oxygen microsensors [17].

Property	sensor type		
	O_2-microelectrode (needle-type)	*O_2-microelectrode (Clark-type)*	*O_2-microoptode*
measuring range (% air saturation)	1 - 500	1 - 500	0.05 - 500 (depending on the type of O_2-sensitive layer)
response time (t_{95})	< 1 second	< 1 second	< 1 second
calibration curve	linear	linear	hyberbolic
	O_2-microelectrode (needle-type)	*O_2-microelectrode (Clark-type)*	*O_2-microoptode*
cross sensitivity	H_2S, pH, ionic strength (especially Ca^{2+}, Mg^{2+})	H_2S	SO_2
longterm stability	poor	moderate	excellent
storage stability	moderate	moderate	excellent (> 1 year with stable calibration curves)
stirring sensitivity	high	moderate	no effect
disturbance of the micromilieu	mechanically, polarisation potential	mechanically	mechanically, excitation light
price	low cost sensor	expensive	low cost sensor

Figure 1 :

a) Schematic of the setup for the measurement of oxygen in soil infiltration water with an optochemical oxygen sensor at a lysimeter station;

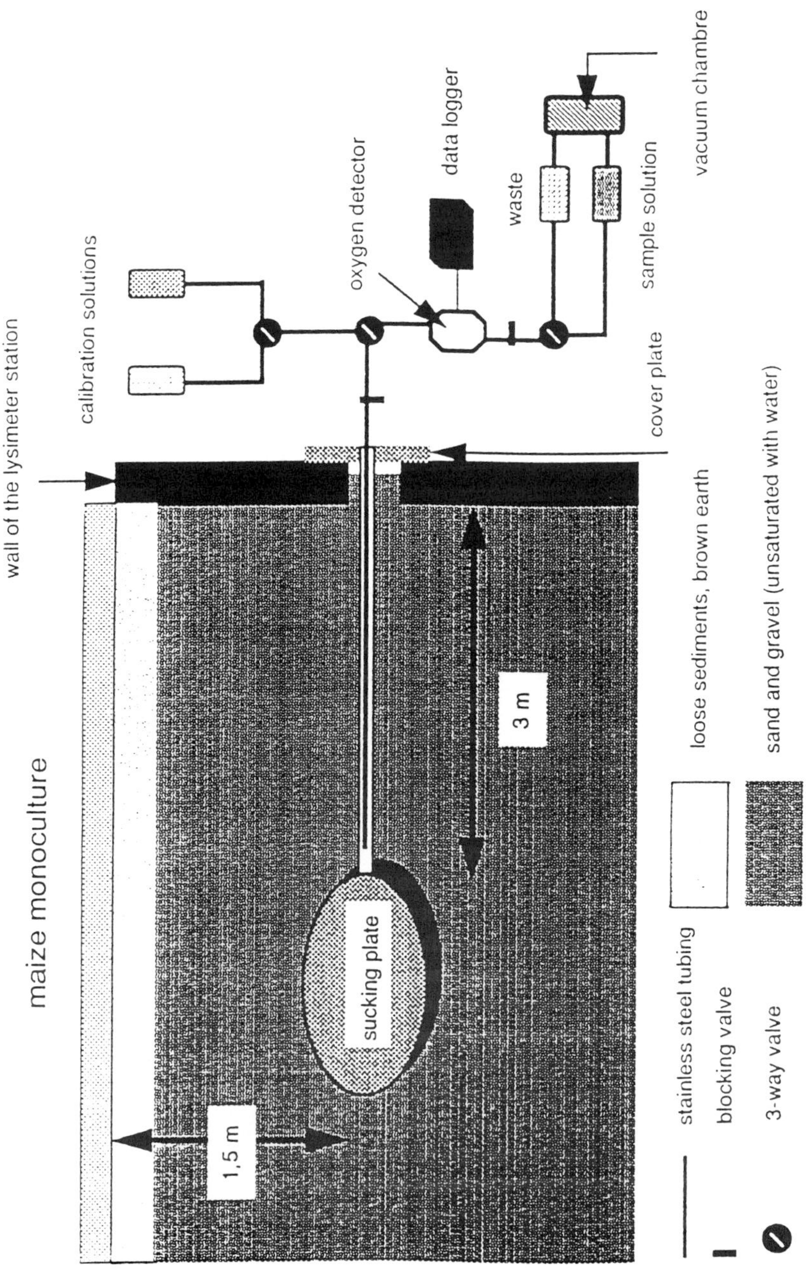

Figure 1 :

b) Comparison of oxygen concentrations in soil infiltration water, which were measured with the optical sensor in the period of March to August 1994, with oxygen concentrations, which were calculated for air-saturated water.

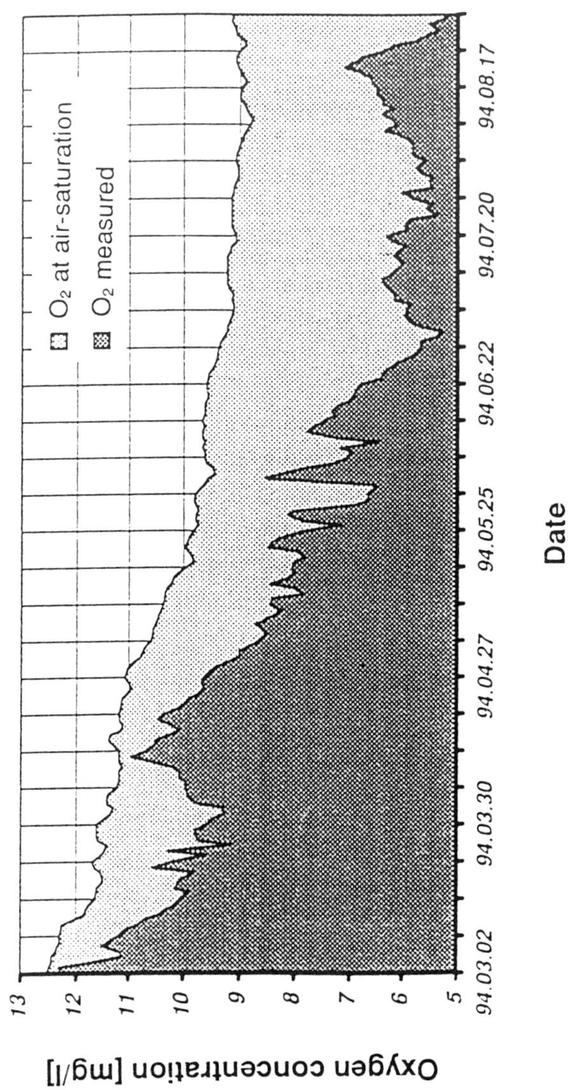

Figure 2 : Oxygen measurement in water with an optochemical oxygen sensor based on a phase measurement technique [15]. Sample solutions were tonometered with calibration gases of varying O_2 concentrations (0, 0.1, 1, 5, 10 and 20.6 % O_2) at 30 °C. The measurement was performed at a flow rate of 0.8 ml/min and at 25 °C.

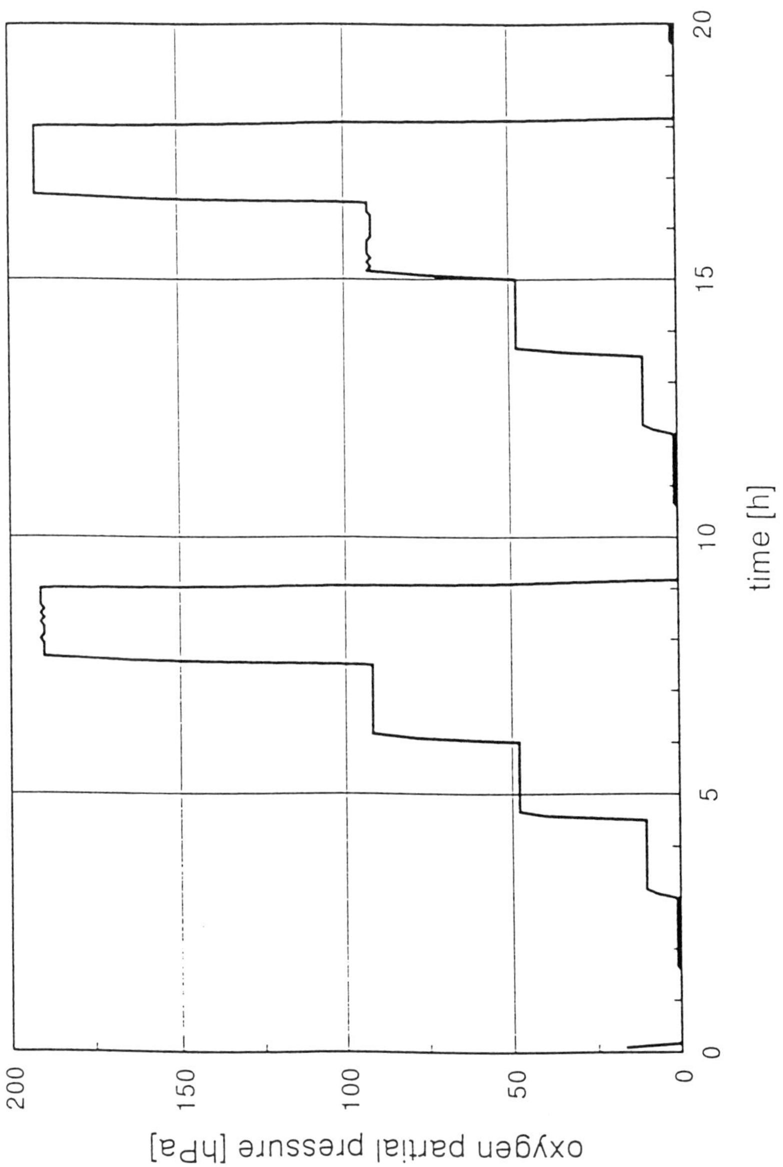

Figure 3 : Schematic of an optode specially designed for oxygen measurements in waste water, groundwater and surface water (outer diameter: 40 mm; length: ca. 190 mm).

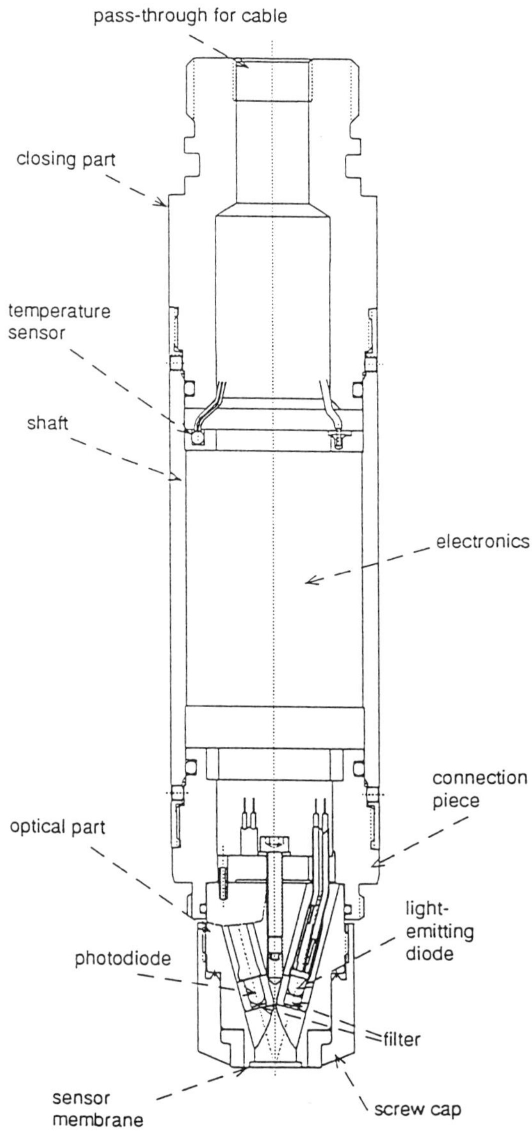

Figure 4 : Schematic of an oxygen microoptode measuring system oxygen profiles measured in a coastal North Sea sediment with an O_2 microoptode and an O_2 microelectrode [16].

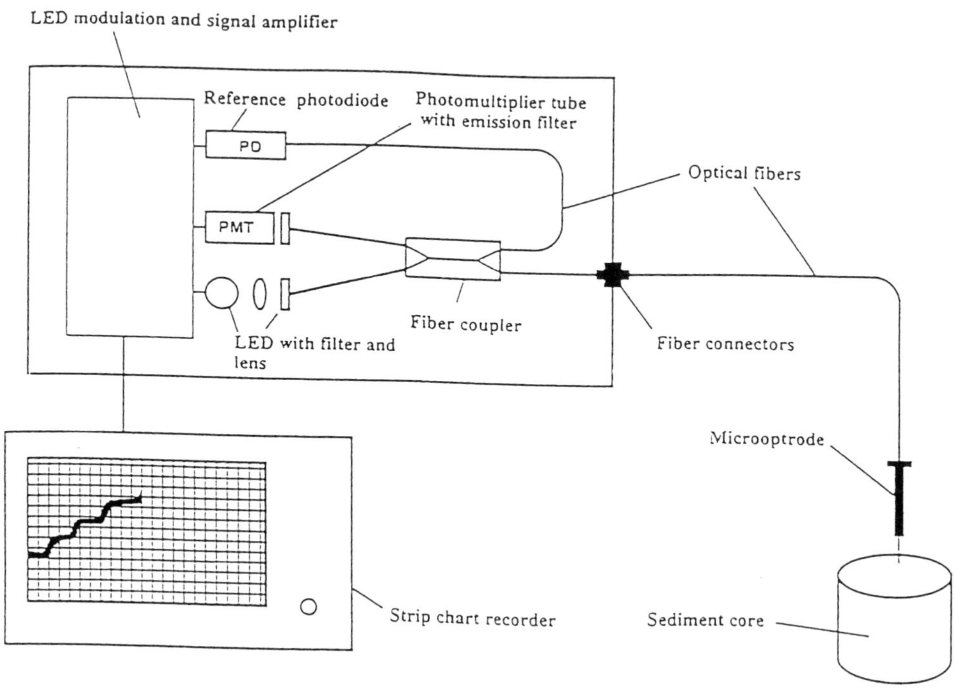

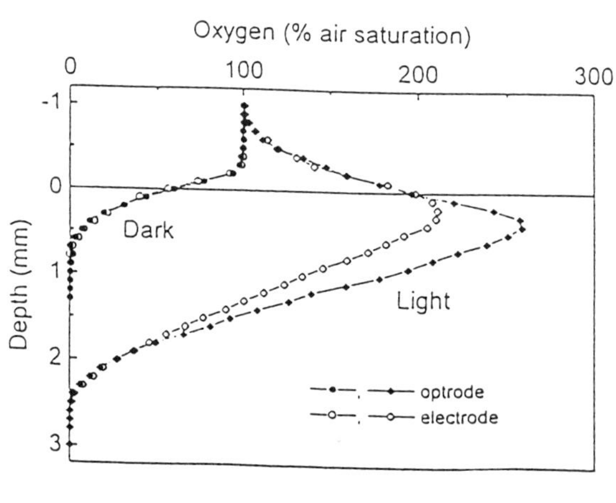

Figure 5 : Absorption spectra of a cellulose-immobilized pH indicator in 0.1 M phosphate buffers of varying pH [18], dependence of the pH calibration curve on the concentration of the buffer employed [19].

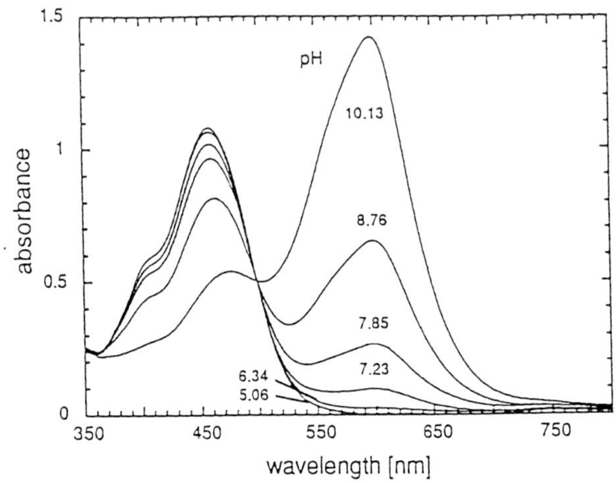

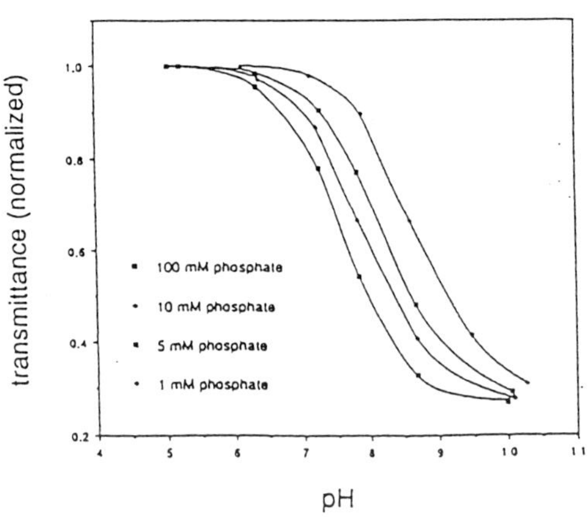

Figure 6 :

a) Carbon dioxide measurement with a fluorescence-based CO_2 sensor in 0.9 % NaCl solution, which was tonometered with gases of varying CO_2 concentration (0, 1, 3, 6, 10 and 15 % CO_2) (sample flow rate: 0.5 ml/min; 37 °C)

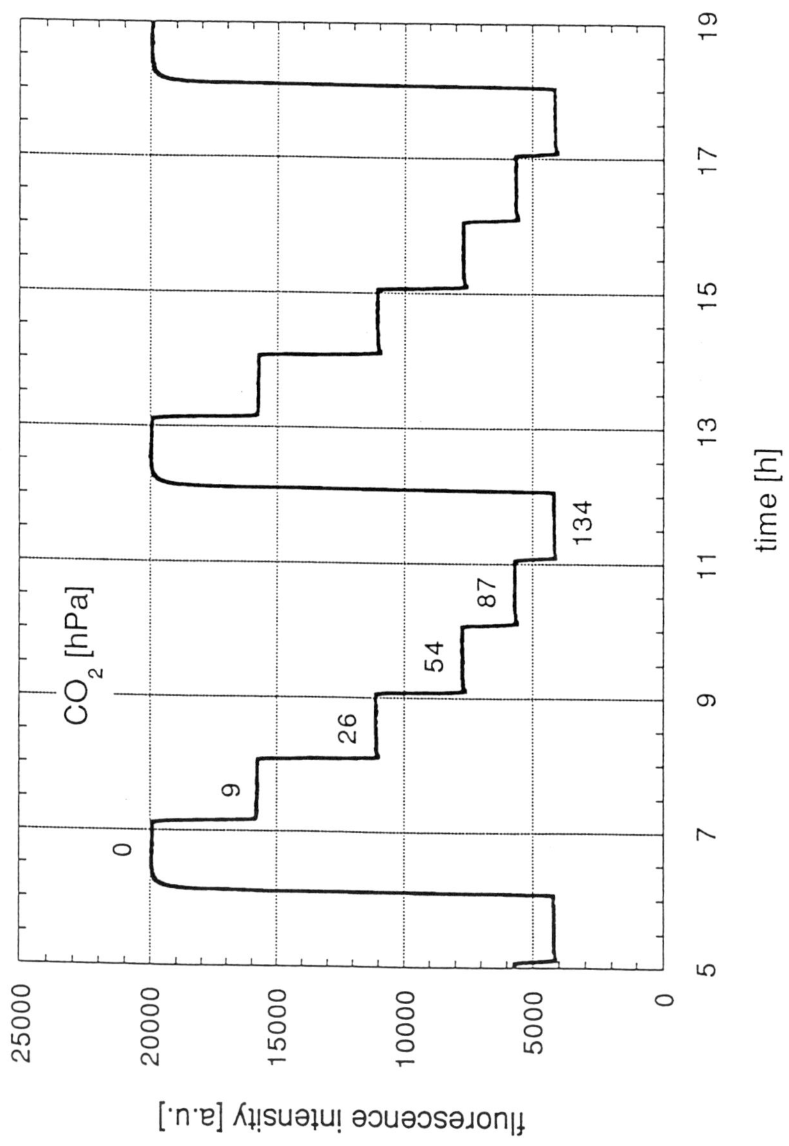

Figure 6 :

b) calibration graph obtained with the same measurement.

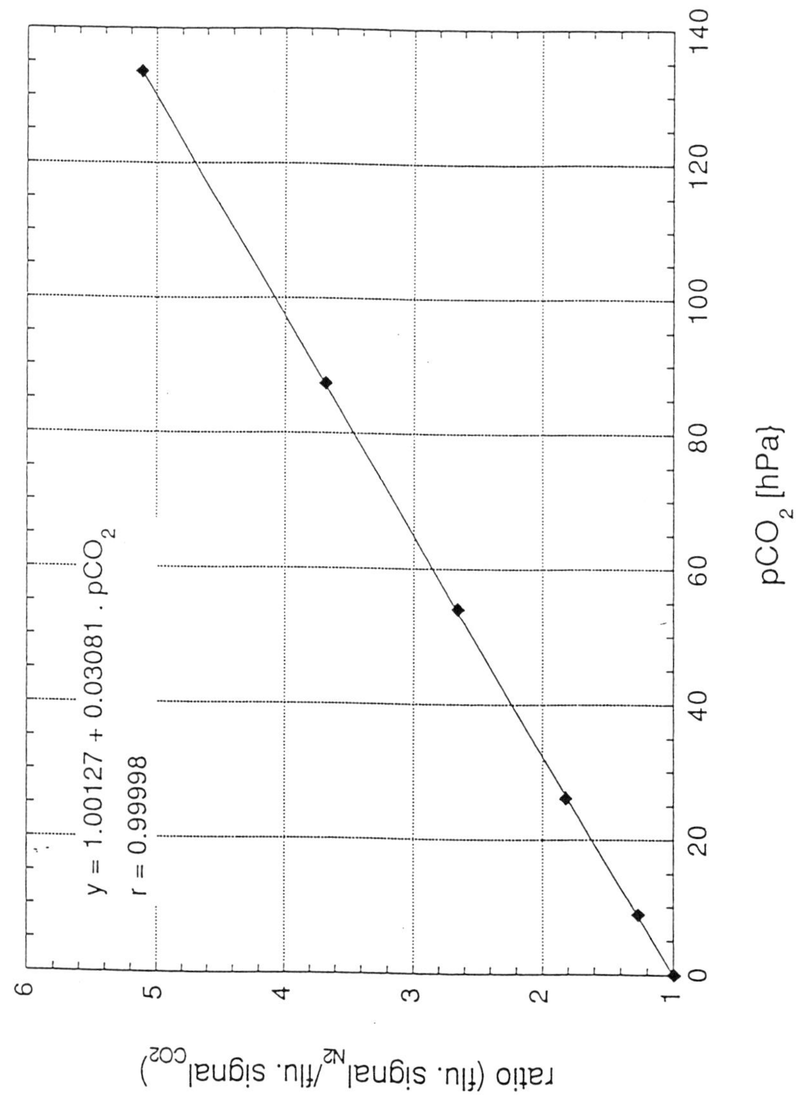

MICROSYSTEMS FOR IN SITU MEASUREMENT OF VOCS IN GROUNDWATERS AND SOILS

J. GUTIÉRREZ, M.C. HORRILLO, J. GETINO, L.ARÉS, J.I ROBLA and I. SAYAGO, Laboratorio de Sensores, Instituto de Física Aplicada (IFA), Consejo Superior de Investigaciones Científicas (CSIC), Serrano, 144, E-28006-Madrid, Spain

1. Abstract

There is a growing interest in the substitution of conventional analytical procedures by suitable selective chemical sensors. Devices allowing continuous monitoring of concentrations in field would be of particular significance in environmental analysis. In this work, a tin oxide multisensor with 15 sensor elements has been prepared using the r.f. reactive sputtering technique. Some components of this multisensor have been doped with Pt, Pd and Al by sputtering. Each sensor has been characterized by electrical measurements for various volatile organic compounds (VOCs) in a concentration range from 50 to 500 ppm in synthetic air at 300 °C.

The best responses have been obtained by Pt and Pd doped sensors mainly for the oxygenated compounds, but for all VOCs tested a good detection is found. For the concentration range used it has been possible to detect the threshold limit values (TLV) for airborne contaminants of toluene, octane and methyl-ethyl-ketone. The influence of the oxygen species (O^- and O_2^-), ionosorbed on the sensing surface, on the VOCs detection is also discussed.

On the other hand, the single component classification and the multicomponent analysis of these VOCs have been carried out. The principal component analysis (PCA) classification permitted to distinguish three families of VOCs. The best results from the multicomponent analysis were obtained for the ILS and PCR methods

2. Introduction

At the present time, the diagnosis of waters, when are contaminated with volatil organic compounds, is based on analytical methods. Up to now, the more used technique is the chromatography in its various forms: TLC (Thin Layer Chromatography), GC (Gas Chromatography) or HPLC (High Performance Liquid Chromatography) and specially HRGC/MS (High Resolution Gas Chromatography/Mass Spectrometry) or HPLC/DAD (High Performance Liquid Chromatography/Diode Array Detector). These techniques allow to identify the compounds not only by means of retention time but also by spectral diferences. In situ measuring techniques are also used for the detection of polluted areas, such as GC/MS for analysis of organic compounds, IR-instruments for PAH, PCP, aliphatic and aromatic hydrocarbons. Immuno-assays and laser sonde are also used [1].

Nowadays, the more important problem related to these analytical methods is that they are expensive and need of a lot of time to carry out the analysis.

Besides, they are not fully adecuated for portable measuring systems. As consequence, few samples are analysed and there is a lack of information for a better evaluation. For these reasons, it is very important to develop faster and cheaper mesuring systems for identification and quantification of hazardous pollutants in contaminated sites. Therefore, these systems must allow to perform continuous and in situ mesurements, for offering real-time data and thus a more effective detection of contaminants.

An alternative to conventional detectors consists of the use of resistive gas sensors, based on metal oxide semiconductors, whose operating principle is based upon the initial reversible reaction of atmospheric oxygen with lattice vacancies in the metal oxide and the concurrent reduction in electron concentration [2]. This reaction produces various oxygen species, at different temperatures and oxygen pressures, which can react with certain gaseous species, as VOCs, producing an electrical conductivity modulation on the metal oxide semiconductor. This type of gas sensor tends to be very sensitive to environmental conditions and often they have very poor ageing characteristics, low selectivity and high cross-sensitivity with other gases or humidity. At present, at the laboratory scale, the use of gas sensor arrays including different metal oxide semiconductors with different additives on a substrate, are getting to overcome the problem of poor specificity using pattern recognition methods (PARC) and neural networks [3].

Few devices can be met on the market nowadays, but none of these commercially available sensors are made with microtechnologies. Planar technologies such as are used in microelectronics can be modified for the fabrication of microsensors. In this way the concept of microtechnological mass production of sensors or sensor arrays can be extended to include inexpensive micro-analysers operating "in situ" with integrated data adquisition and evaluation.

In this work, a new field measuring system is presented, formed by the multisensor device on alumina substrate, the electrical heater, the test chamber and the electrical connector (Fig.1). An electrical heater is used to achieve the operating temperature of the sensor. It can be powered up to 12 V dc, for temperatures up to 500 ºC.
The heater is placed within an isolating ceramic material. The test chamber is made in stainless steel and it has an effective volume of 20 cc. At the top of the chamber the gas inlet and outlet are located. At the bottom of the chamber the multisensor device is located.
The electrical connectors are made in platinum ending in a DB25 connector. The data adquisition system consists of a digital multimeter (DMM) with a scanner card up to 16 channels, a personal computer with an IEEE card and software for data storing and treatment.

The VOCs extraction from soils or groundwaters and their transport to the field measuring system can be performed by permeation cells based on membranes. It is known that elastomeric membranes interact preferencially with organic compounds, therefore they have high selectivities for organics over water. Commercial membranes as: PDMS (Polydimethylsiloxane), PEBA (Polyetherblock-poliamides), SPC (Silicone-polycarbonate copolymer) and PUR (Polyurethane) used in multicomponent pervaporation allow a good separation of VOCs. It has been shown that the solubility of penetrants in the elastomeric membranes determines the membrane permeability and selectivity to a large extent, therefore the hydrophobicity of organics determines the sequence of the permeability and the separation factor [4].

The superposition of membranes onto the sensor, using a appropriate assembly, joined to the heating facilities of the sensor, up to 500Cº, allows to eliminate the distortion in measurement due to the water vapour presence, however in case of direct application on waters, a deeper study about membranes must be carried out.

The results obtained from this work will be used as reference measurements for future "in situ" experiments on gases extracted from groundwaters through permeation cells with specific membranes and transferred to the field measuring system. The vectorial or matricial response is considered as a characteristic "fingerprint" of the contaminants.

3. Experimental

The multisensor is designed with 15 sensor elements distributed on circular sectors on an alumina sheet of one inch diameter as substrate. The base material was tin oxide and different dopants were used, such as: Al, Pt, Pd. The tin oxide thin films (200 nm of total thickness) were grown by reactive sputtering at 250 ºC using a SnO_2 target under 10/90 oxygen/argon mixture at a total pressure of 0.5 Pa. Each doped element was prepared with different amounts of dopant, introduced as an intermediate discontinuous layer between two layers of tin oxide each one with the same thickness (sandwich structure), changing the deposition time during the sputtering process. Table 1 shows the multisensor distribution. It was thermally treated in air at 520 ºC during 4 h for controlling the material morphology (stoichiometry and grain size of the oxide, and dopant distribution) and the semiconductor resistance stability since the parameters of preparation and annealing are fundamental in order to obtain a good detection.

Table 1. Multisensor distribution

Sensor No.	Semiconductor Material	Dopant (time)
1	SnO_2 (1 layer - 180 nm.)	-
2	SnO_2 (1 layer - 180 nm.)	-
3	SnO_2 (1 layer - 180 nm.)	-
4	SnO_2 (1 layer - 180 nm.)	-
5	SnO_2 (2 layers - 90 nm. Each)	Al (16 s)
6	SnO_2 (2 layers - 90 nm. Each)	Al (12 s)
7	SnO_2 (2 layers - 90 nm. Each)	Al (8 s)
8	SnO_2 (2 layers - 90 nm. Each)	Al (4 s)
9	SnO_2 (2 layers - 90 nm. each)	Pt (16 s)
10	SnO_2 (2 layers - 90 nm. each)	Pt (12 s)
11	SnO_2 (2 layers - 90 nm. each)	Pt (8 s)
12	SnO_2 (2 layers - 90 nm. each)	Pd (16 s)
13	SnO_2 (2 layers - 90 nm. each)	Pd (12 s)
14	SnO_2 (2 layers - 90 nm. each)	Pd (8 s)
15	SnO_2 (2 layers - 90 nm. each)	Pd (4 s)

The multisensor was placed in a stainless steel test chamber and the resistance measurements were carried out at 300 ºC with a constant flow rate of 200 ml/min at VOCs concentrations (benzene, toluene, propanal, chloroform, MEK, and octane) varying from 50 to 500 ppm in synthetic air. These mixtures were prepared by means of mass flow controllers. The exposure time to each tested gas concentration was 20 minutes. The measuring (resistence and temperature) and mixing processes were automatically carried out by means of DMMs and PCs.

On the other hand it has been used a quadrupole mass spectrometer BALZERS QMA 130 with working range from 1 to 300 amu provided with a special device for gas analysis at atmospheric pressure, so that the desorbing products coming from VOCs tested could be analysed. The measurements with the mass spectrometer were performed with the analysis probe placed some milimeters above the multisensor, when the gas mixture (VOCs/air) was interactioning with the multisensor surface, at 300ºC and for 500 ppm of each VOC.

With regard to the single component classification of the sensor array it has been performed to the six VOCs tested. The concentration of each VOC was varied from 50 to 500 ppm/air per experiment. The experiments were repeated between three and ten times for each VOC. The responses were measured after 20 minutes of each gas exposure.

The multicomponent characterization has been performed upon four VOCs (benzene, toluene, octane and propanal) gas mixtures. The selected number of observations was 81, corresponding to the mixtures of four gases and three possible concentrations per gas (0, 50, 100 ppm/air). In this case the responses were measured after 10 minutes of gas mixture exposure.

4. Results and discussion

4. 1. Sensitivity and reactivity

The sensitivity values (S) of the multisensor, defined as the ratio between the semiconductor conductance change and the semiconductor conductance in air, i.e. $S=(G_{gas}-G_{air})/G_{air}$ to 500 ppm for benzene, toluene, chloroform, octane, propanal, and MEK (methyl-ethyl-ketone) at 300 ºC are shown in Fig. 2. The best detection responses (higher sensitivity values) were obtained for the compounds with the carbonil group (aldehydes and ketones) in comparison with those obtained for aromatic hydrocarbons (benzene and toluene) and for chlorinated compounds, as chloroform, which showed the lowest sensitivity values.

A good sensitivity value for octane also appears at this temperature. Since these gases are reducing, the sensor elements doped with Pt and Pd showed the best response, because such dopants activate the oxidation reactions produced on the semiconductor surface.

At 300 ºC, Pd is oxidized, and its catalytic behaviour could be the same as Pt, i.e. by mass transfer (spillover) of the gas from the catalyst nanograins to the semiconductor nanograins where oxygen species are ionosorbed. Furthermore, it was a common feature that the sensitivity increased when the doping level decreased for both dopants for the thicknesses of tin oxide film indicated.

The sensitivity results only were slightly better for Al-doped sensors than for undoped, because this dopant is suitable for oxidant gases. According to the order of better sensitivity values, the sequence of reactivity would be :

propanal > MEK > octane > toluene > benzene > chloroform

At the temperature and concentations tested, it has been possible to measure the TLV for airborne contaminants of toluene, MEK, and octane (100 ppm, 200 ppm and 300 ppm, respectively) as it is seen in Fig. 3a, 3b and 3c. It can also be seen in such figures that the resistance values have a higher decrease when the gas concentration increases and that the reaction equilibrium is reached more quickly for higher concentrations than for the lower ones, which demostrates that the response rate depends on the gas concentration. These facts are more significants for Pt doped sensors.
In order to study the response behaviour of this type of gas sensors, based on the electrical conductivity of gas sensors modulated by the presence of gases, in this case VOCs, and to relate the multisensor outputs to the gas concentrations, it is important to take into account that the reaction processes take place on the semiconductor surface. It is well-known that these sensors work through chemisorption and/or oxidation-reduction reactions. Furthermore, during decomposition and oxidation of carbon-containing molecules, common intermediate compounds are produced on the SnO_2 surface, in such a way, that VOCs molecules exposed on the sensor surface, could produce, mainly as intermediate, oxidized species, radicals, OH groups and adsorbed hydrogen from the reaction steps.

The semiconductor oxide superficial conductivity depends initially on the oxygen vacancies formed in the preparation process, but such conductivity depends later on the adsorbing molecules, it depends on the density of donors (adsorbed hydrogen, oxygen vacancies and reducing molecules) and acceptors (chemisorbed oxygen species and oxidant molecules). The most important aspect to study is the reaction of oxygen species, ionosorbed on the tin oxide surface (O_2^-, O^-), with these volatile organic compounds. Such oxygen species are produced while the thermal treatment is carried out and at the working temperature of the multisensor. It is important to take into account that the reactivity of the oxygen species is as follows : $O^- > O^-_3 > O_2^-$, and with regard to oxidation reactions, these oxygen species adsorbed are classified as "electrophylic" reactants which in polycrystalline materials, like tin oxide, preferentially attack the C-C bond [5].

According to the results obtained, from multisensor responses, it would be possible to have intermediate products, as those previously mentioned, and desorbing products, such as: propanoic acid or propanol from propanal, acetaldehyde and ethane from oxidative scission of MEK, since the ketones show more difficulty to be oxidized, and different oxygenated compounds and/or carbon dioxide and water from octane. The lower sensitivity values obtained for benzene, toluene and chloroform could be due to the absence of oxidation reactions between these compounds and the adsorbed oxygen species. With regard to results obtained from the measurements performed with the mass spectrometer [6], it has been possible to observe 2-propanol formed from propanal, CO_2 and H_2O from octane and CO_2, H_2O and acetaldehyde were produced from the scission of MEK (Fig. 4). For the other VOCs, no new compounds were observed in the spectra.

The response has a different behaviour for the different VOCs tested: proportional to the square root of the concentration for aromatic and chlorinated compounds and almost linear, although with change slope on increasing the concentration, for oxygenated and aliphatic compounds [6].

4. 2. Classification

With the data from 94 observations of the single-component characterization, it has been performed PCA to obtain the classification of the six VOCs. The tin oxide sensor signals (resistances), Y_{ij}, for only one gas component j present in air can be approximated by [7]:

$$Y_{ij} = R_{oi}\,(1 + B_j\; X_j\,)^{c_{ij}} \quad (1)$$

where R_{oi} is the resistance of the sensor element i in air, X_j is the concentration of the test gas j and B_j, c_{ij} are model parameters depending on temperature and tin oxide surface.
To eliminate the gas concentration influence on the sensor signals, these are normalized considering B_j as constant:

$$p_{ij} = \frac{c_{ij}}{\sum_{i=1}^{p} c_{ij}} \quad (2)$$

The resulting concentration-independent signals, p_{ij}, are mean-centred and variance-scaled and then used as the inputs for PCA [8]. The optimum system rank was 3, since higher rank vectors represented only noise. Considering the loadings of the first two factors, three groups of sensors could be separated: group 1: sensors 1-10, group 2: sensor 11, group 3: sensors 12-15.
The sensors in each group were redundant and indicate collinearity in the response matrix. Considering the loadings of factor 3, two subgroups were found from the previous group 1: Subgroup 1: sensors 1, 2, 4 and subgroup 2: sensors 3, 5-10.

PCA was performed over the response matrix considering three VOCs families as classes: oxygenated (propanal, MEK), chlorinated (chloroform), and aromatic/aliphatic (benzene, toluene/octane). In Fig.5, the results from PCA are shown. The scores of the three families are plotted for factor 2 vs factor 1 (Fig. 5a) and factor 3 vs factor 1 (Fig. 5b).

From only the first two factors, the chlorinated family can be separated quite well, but the other families overlap. Considering factor 3, a more efficient discrimination was found [9].

4. 3. Multicomponent analysis

With the measured data from the multicomponent characterization it has been carried out the multicomponent analysis in order to obtain the quantitative determination of the components of a VOCs mixture. The multicomponent analysis has been accomplished in two steps: the

linearization and the multivariate linear regresion. This has been performed according to the Partial Model Building (PMB) method [7].
With the linearized data set, linear regresion methods, like CLS, ILS, PCR, and PLS, were applied obtaining four regression matrices, one per method. The quality of applied multivariate regression models were not very good. The coefficients of multiple determination were far from one, indicating that the residual sums of squares were high. The means values of the relative prediction errors in percent and the mean values of the width of the 95% prediction intervals in ppm are shown in Table 2. In general, the lowest errors were obtained for PCR and ILS methods and therefore the best analysis of VOCs mixtures. The width of the ILS prediction intervals was narrower than the width of PCR intervals, but the ILS relative prediction errors were higher than PCR relative errors (except octane). With more data, the errors of the analysis would be minimized and a better classification could be obtained [9].

Table 2. Mean values of relative prediction errors and width of prediction intervals for MVA methods

Method	Relative Prediction Errors (%)				Width of Prediction Intervals(ppm)			
	Benzene	Toluene	Octane	Propanal	Benzene	Toluene	Octane	Propanal
CLS	428.4	43.0	27.9	47.4	817	88	50	235
ILS	60.6	45.8	20.9	43.4	146	70	20	54
PCR	34.6	29.8	25.7	30.2	131	108	33	120
PLS	45.0	34.7	19.4	39.0	124	113	63	119

5. Conclusions

It has been possible to measure low VOCs concentrations (detecting some TLVs) by means of the multisensor system built. The best sensitivity values corresponded to the most reactive compounds, following this order:

oxygenated > aliphatic > aromatic > chlorinated

Pt and Pd were good dopants for VOCs detection. The amount and the good distribution of the dopant on the semiconductor particles are very important to improve the detection.

About the classification of these compounds, considering three principal components it was possible to separate well the chlorinated family from the other ones. Between the oxygenated and aromatic & aliphatic there was overlapping. The best analysis of VOCs mixtures were obtained with ILS and PCR methods.
The width of the ILS prediction intervals was narrower than the PCR intervals, but the ILS relative prediction errors were higher than the PCR relative errors (with the exception of octane).

Departing from the present results after using the semiconductor sensor array, our challenge at the present consists of the integration of these sensors on silicon substrates towards the development of microsensor arrays due to the compatibility between the microfabrication technology (sputtering) and the microelectronic technology. Microsensors integrated present sustancial advantages respect the conventional sensors on ceramic substrates until now developed, such as smaller size, lower heating power, accurate and independent temperature control and eventual lower cost based on batch production. Besides the electronic and the sensor can be fabricated on one single chip.

The development of integrated gas multisensors associated with microelectronic technology and data processing, will allow the realization of analytic microsystems that can operate "in situ". Unfortunately until now, developments concerning integrated gas sensors on silicon are relatively scarce. Our preliminary studies and results of a micromachined solid state integrated sensor can be seen in ref. 10. At present we are working in the development phase.

6. Acknowledgements

This work has been supported by the European Commission under Environment Programme Project EV5V-CT94-0546 VOCINSOL ("Measurement and modelling of VOC mobility in soils and groundwaters for environmental risk assessment") and by the Comisión Interministerial de Ciencia y Tecnología (CICYT) under a National Programme Project: MIGAS: ("Microsensors integrated in thermally isolated structures"), in colaboration with the Centro Nacional de Microelectrónica (Barcelona)

REFERENCES

[1] M. Acilu, Contaminated Soil'95, Vol. 1, W.J. Van den Brink, R. Bosman, F. Arendt, Editors, p. 481 (1995)
[2] M.C. Horrillo, J. Gutiérrez, L. Ares, J. Robla, I. Sayago, J. Getino, J. Agapito, The influence of the tin oxide deposition technique on the sensitivity to CO, Sensors and Actuators, B, 25, (1995) 507-511
[3] M. Homberg, F. Winquist, I. Lundstrom, J.W. Gardner, Identification of paper using hybrid electronic noses, Sensors and Actuators, B, 26-27, (1995) 246-249
[4] J. Wenchang, K. Subhas, S.K. Sikdar, Modelling of multicomponent pervaporation for removal volatil organic compounds on water, Journal of Membrane Science, 93, (1994) 1-19
[5] J. Haber, Proc.of the 8th International Congress on Catalysis, Berlin,Verlag Chemie: Weinheim (FRG), Vol. 1, pp.85.
[6] M.C. Horrillo, J. Getino, J. Gutiérrez, L. Arés, J. Robla, C. García, I. Sayago, Measurements of VOC's in soils through a tin oxide multisensor system, Proc. Eurosensors X, Leuven, Sept. 8-11, (1996), Vol.2, p.669 (Accepted in Sensors and Actuators, B, 1997)
[7] G. Horner and C. Hierold, Gas analysis by partial model building, Sensors and Actuators, B, 26-27, (1995) 246-249

[8] E.R. Malinowski, Factor analysis in chemistry, J.Wiley, New York, (1991)
[9] J. Getino, M.C. Horrillo, J. Gutiérrez, L. Arés, J. Robla, C. García, I. Sayago, Analysis of VOCs with a tin oxide sensor array, Proc. Eurosensors X, Leuven, Sept. 8-11, (1996), Vol.2, p. 673 (Accepted in Sensors and Actuators, B, 1997)
[10] A. Götz, I. Gràcia, M.C. Horrillo, J. Getino, C. García, J.Gutierrez, A micromachined solid state integrated gas sensor for detection of aromatic hydrocarbons, Proc. Eurosensors X, Leuven, Sept. 8-11, (1996), Vol.2, p.549 (Accepted in Sensors and Actuators, B, 1997)

Figure 1 : View of the measuring system.

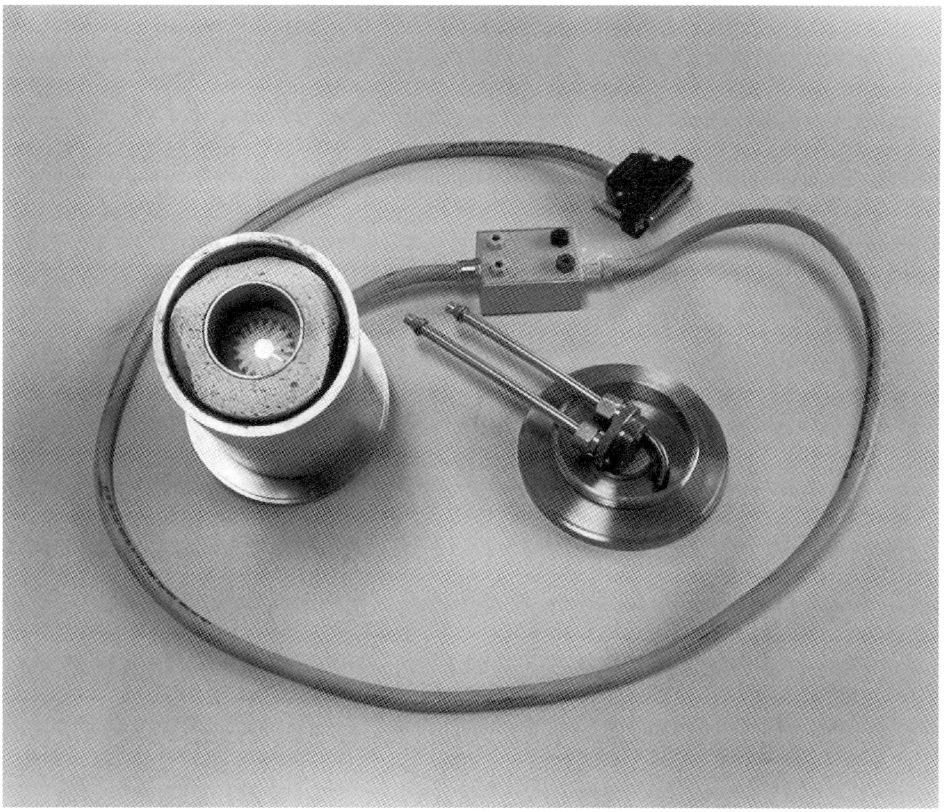

Figure 2 : Sensitivities to 500 ppm in air for VOCs at 300°C of a doped tin oxide multisensor.

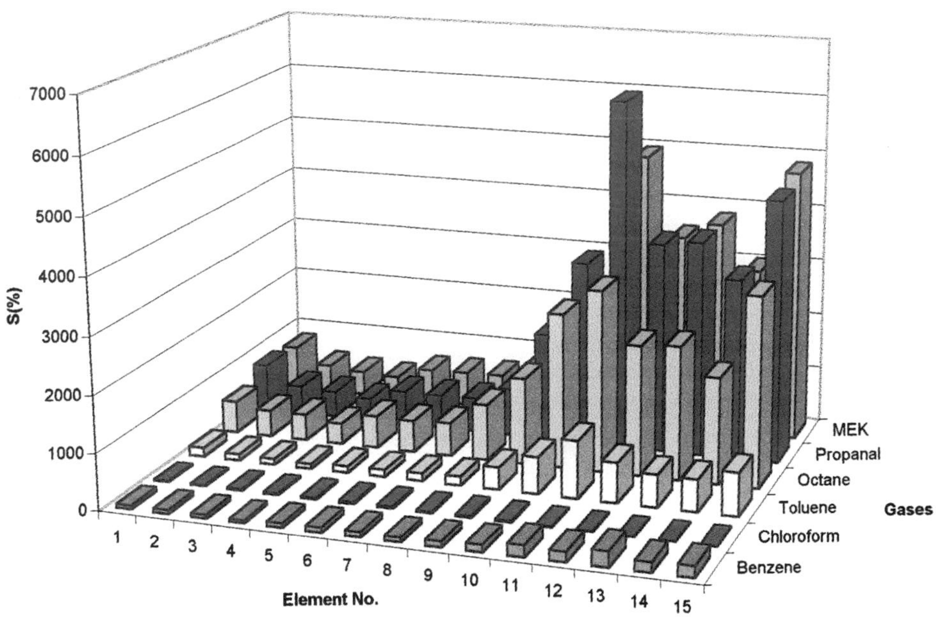

Figure 3 : Responses to a) toluene and b) MEK for an undoped sensor (S1) and Pt doped sensor (S11).

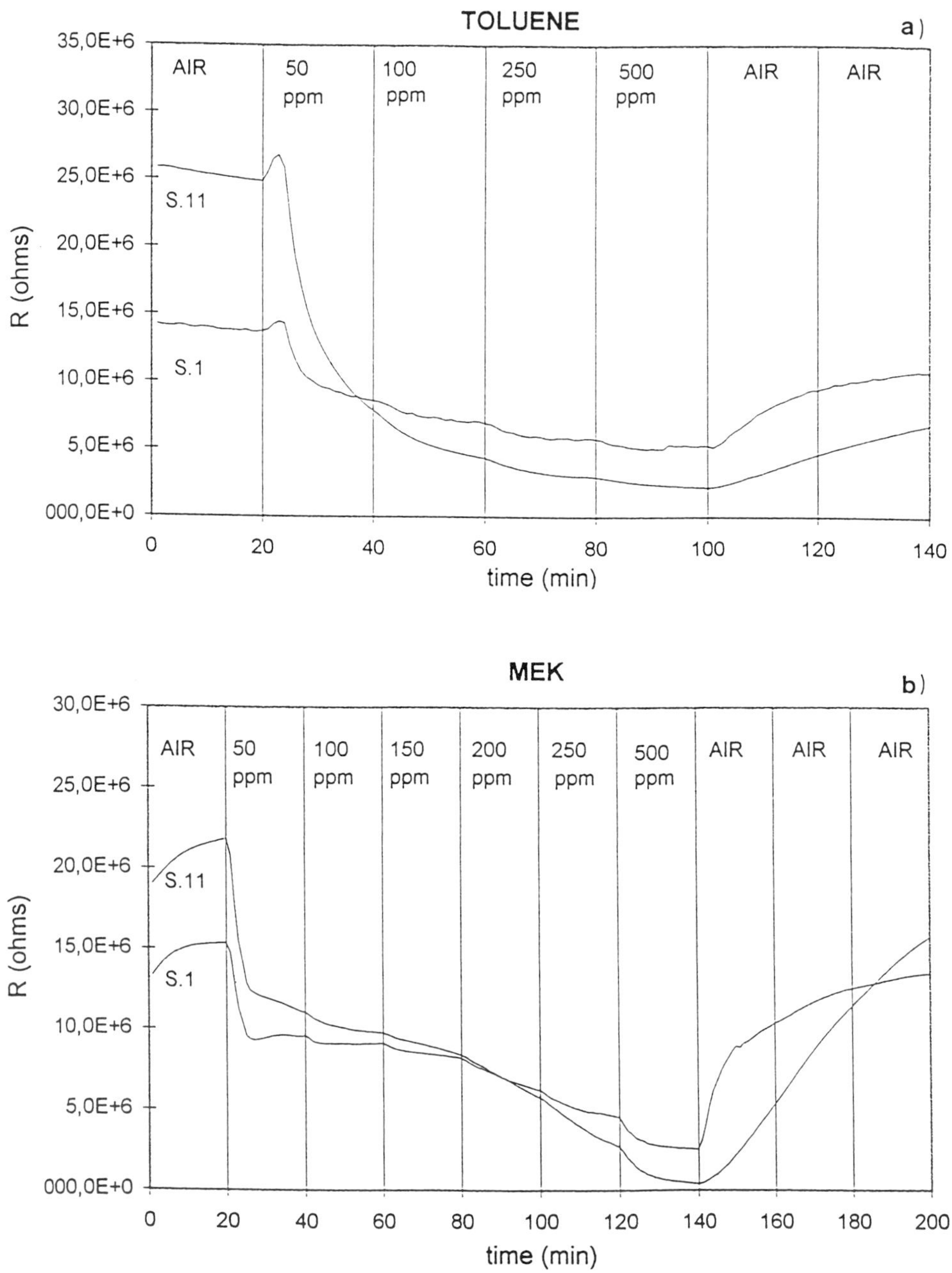

Figure 3 : Responses to octane for an undoped sensor (S1) and Pt doped sensor (S11).

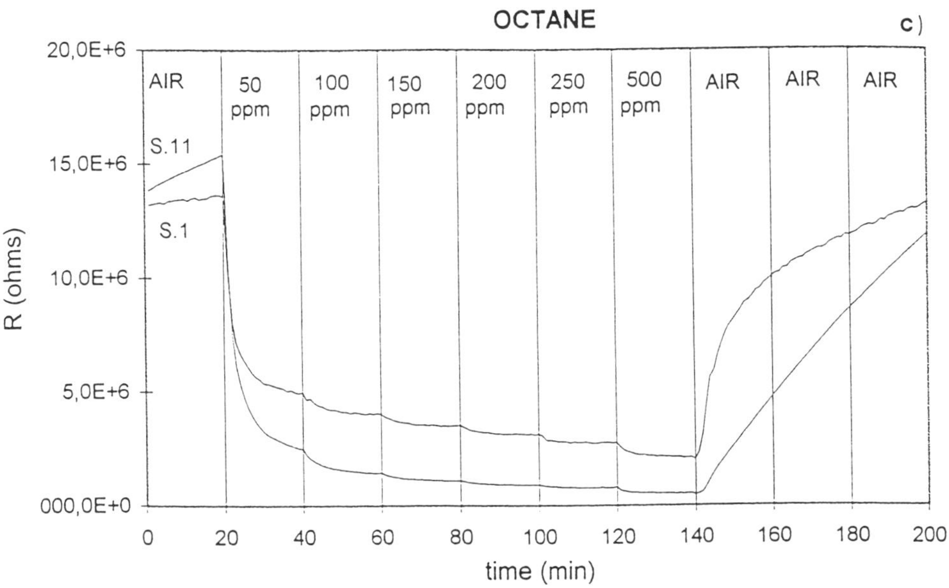

Figure 4 : Mass spectrum of 500 ppm of MEK interactioning with the multisensor.

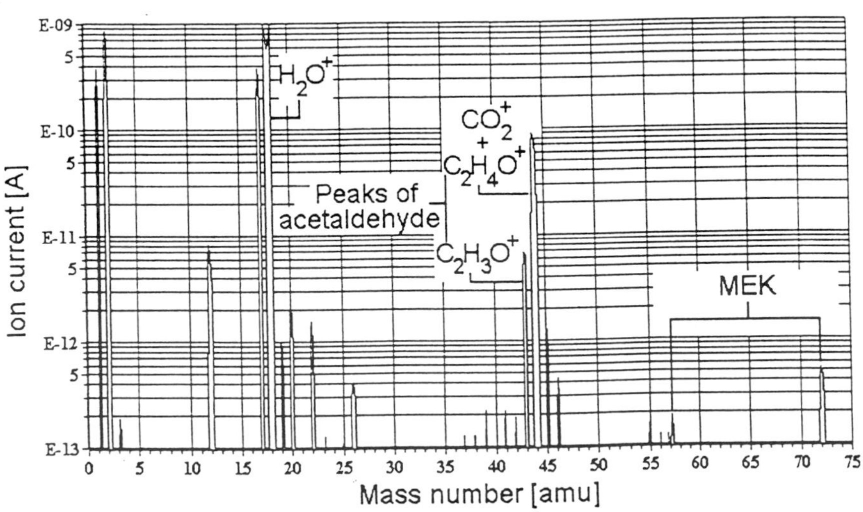

Figure 5 : Results of PCA for the 15 element sensor array showing the scores of three VOCs families for principal components a) 1 and 2, and b) 1 and 3.

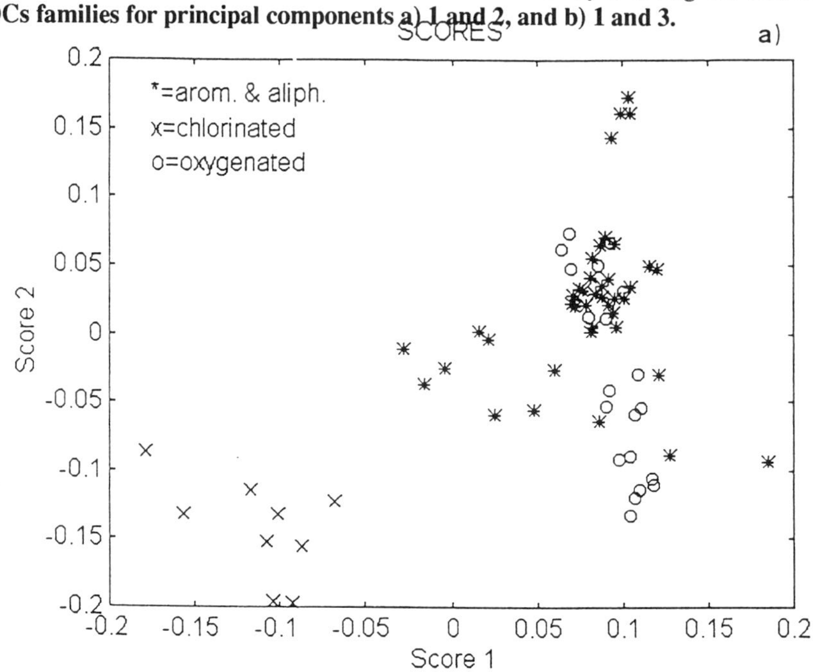

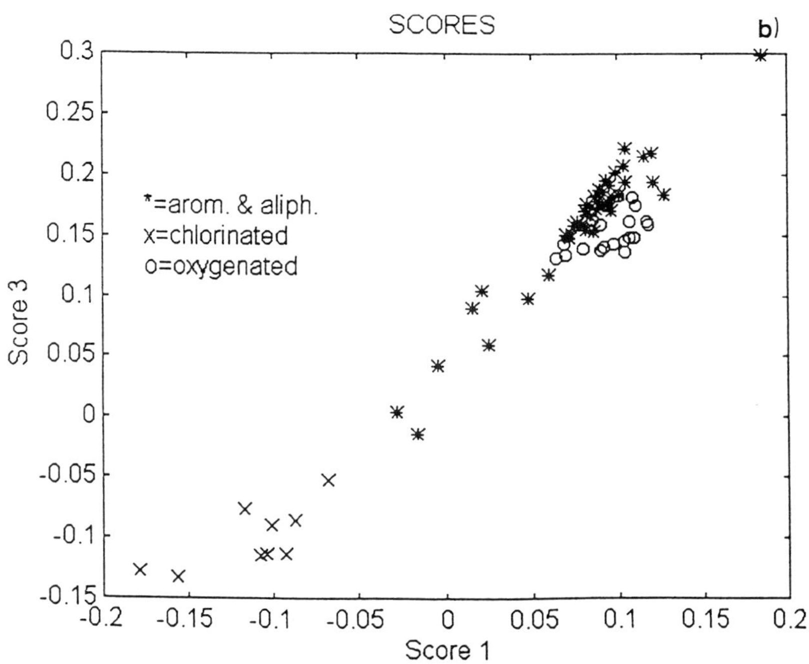

HEADSPACE ANALYSIS USING CONDUCTING POLYMER SENSOR ARRAYS

Peter A PAYNE, K. C PERSAUD, M E HASSAN AMRANI and G. KING, Department of Instrumentation and Analytical Science, UMIST, Manchester, United Kingdom.*
**AromaScan plc, Crewe, Cheshire, United Kingdom.*

1. Abstract

Research efforts have been under way for over twenty years aimed at investigating and understanding the use of conducting polymers to monitor volatile chemicals. In some cases, this work has led to commercial exploitation. The mechanisms by which interaction between the polymer surface and the volatile chemicals give rise to electrical changes in the polymer are not fully understood and this paper will briefly review this fundamental topic. It will go on to show that major advantages accrue from interrogating conducting polymers using high frequency techniques rather than the dc methods used by most researchers. These advantages will be quantified by reference to simple lumped circuit models that have been built for a sensor configuration in use by the present authors. In addition, comparisons between the dielectric properties of the sensors predicted by the model and from a number of measurements will be used to justify the model structure. The paper will also discuss conducting polymer array technology, methods of interrogating such sensor arrays and their deployment as part of new sampling strategies aimed at eventually achieving ppt levels of detection of contamination in water using headspace analysis.

2. Introduction

Increasingly, stringent regulations are being put in place concerning the trace levels of undesirable and, indeed, dangerous compounds that can exist in potable water. Despite this, water companies frequently experience customer complaints, often concerning off-odours and off-tastes present in water supplies. The problem is particularly difficult to deal with since the human nose is exquisitely sensitive to certain of the potential contaminants of drinking water. A good example of this is geosmin, for which a typical sensitivity for the human nose is below 10 parts per trillion (10 parts in 10^{18} v/v) (ppt). There is no instrumental method of directly measuring geosmin at this concentration, even though in the headspace above water contaminated at 10 ppt we are likely to find much higher concentrations. This is because the geosmin molecule prefers not to be in the presence of compounds such as water. Headspace analysis is, therefore, a potentially good approach for this type of compound and when combined with the appropriate sampling techniques, conducting polymer sensor arrays are able to detect such compounds at ppb levels. Further work is still required to get down to ppt concentrations.

3. Gas, odour and aroma sensing

We can approach the gas sensing problem in two ways: if we know the gas that is to be detected and furthermore, that there is little risk of contamination from other species, then a variety of technologies exist that can be made specific for such an approach. Examples of such technologies are: chemiresistors constructed from metal oxides or phthalocyanines, potentiometric and amperometric sensors employing electrochemical techniques, and optical and infrared absorption sensors. These technologies are capable of dealing with gases such as NO_x, H_2, NH_3, CO, CO_2 and CH_4 [1]. Another technology of use in detecting combustible gases is that based on catalytic gate MOSFETs. There are also examples of the use of pellistors as catalytic metal sensors and these are suitable for detecting a wide range of organics. In the main, these technologies are used in the form of single sensors.

The alternative approach is to use arrays of sensors. The first report of the use of sensor arrays appeared in 1982 [2]. The authors showed that by combining a set of commercially available metal oxide sensors, deliberately selected to have non-specific response to a wide range of gases and odours, the response obtained from such an array had some of the characteristics associated with the mammalian olfactory system. Since the publication of this paper, enormous effort has been directed into understanding the characteristics of sensor arrays constructed from a number of technologies, such as conducting polymers, quartz crystals and surface acoustic wave based devices. Many of these are also now available commercially and there is even the promise of commercially available mixed technology arrays, although to date, there is little evidence that this approach attracts much in the way of advantage.

3. 1. Conducting polymer arrays: construction and characterisation

There has been considerable interest in electrically conducting polymer materials for many years. The first report on the chemical oxidation of pyrrole appeared in 1888 and subsequently reports on the polymerisation in the presence of hydrogen peroxide to yield what was commonly referred to as pyrrole black had to wait until 1916. It was not until 1968 [3] that the first report of electrochemical production of polypyrrole appeared. Since then this material has been intensively studied and there is a considerable body of literature, see for example Skotheim [4]. In the 1970s reports began to appear on the gas detecting properties of certain conducting polymers [5].

We have been involved for some time in studies of the chemistry and applications of organic conducting polymers for sensing of volatile chemicals. The strategy adopted has been to develop sensing systems where individual elements in an array show broad, and overlapping selectivities to various chemical species, each sensor element responding more selectively to certain chemicals. This approach has the advantage that the array can respond to many thousands of chemical species due to the broad selectivity of the adsorbent surfaces. On the other hand, extremely selective information for discrimination between adsorbed chemical species or mixtures can be obtained by analysis of the cross-sensitivities between sensor elements. The relative responses between sensor elements produce patterns that may be unique "fingerprints" and that may be used as odour descriptors.

Polymers of some heterocyclic materials show reversible changes in electrical conductance with adsorption and subsequent desorption of odour molecules. There is high sensitivity to polar molecules, and although chemical selectivity of individual polymers is broad, arrays of sensors that contain different functional groups on the surface show different but overlapping responses to different chemical families. Functionalisation of conducting polymers and their applications have recently been reviewed [6]. There is a quantitative relationship to odour concentration since the change in conductance measured is directly proportional to the concentration of adsorbed odour molecules and their binding affinity to the conducting polymers [7]. Of advantage in these kinds of sensing materials is the ability to react rapidly and reversibly at ambient temperatures, and that they are more sensitive to polar volatiles than to non-polar volatiles. This correlates well with the range of volatile chemicals to which the human nose is most sensitive, and therefore in many cases the sensor responses can be correlated to human perception.

3. 2. Mechanisms of interaction between polymer surface and volatile chemicals

The molecular recognition capabilities of organic semi-conducting polymers means that they are materials with very powerful sensing properties. Specific stimuli (ie, gaseous environments) are recognised by characteristic changes in resistance or conductivity of the material following adsorption of the vapours at the polymer surface. Conducting polymers have, therefore, been described as "intelligent materials" [8].

A conducting polymer gas sensor may be prepared electrochemically or chemically. However, both methods produce materials that are oxidatively doped, and generate a delocalised positive charge along the polymer backbone (ca. one charge per four monomer units). This positive charge means that an anionic species (counter-ion) must also be present within the polymer matrix to retain electroneutrality and stability within the system (Figure 1). It is the presence of these charges along the polymer backbone that facilitates molecular interaction.

There are a variety of interactions that may occur at the sensor surface between the polymer and the vapour molecules, such as ionic-ionic or ionic-dipole [9]. However, all of these interactions create physical changes in the polymer structure. A water molecule for example may interact with polypyrrole via the lone pair of electrons on the water oxygen atom and the positive charge on the polymer chain (dipole-ionic). The interaction at the positive sites therefore means that the conductivity of the polymer is reduced. Alternatively, interaction with electrophilic gases (where electron density is withdrawn from the polymer) results in an increase of charge carriers and therefore also an increase in conductivity [10]. The chemical interaction that occurs at the polymer surface may also be influenced by choosing the appropriate monomer and counter-ion contribution [8]. By altering the counter-anion within the matrix it is possible to vary the gas sensing capability, since anion-exchange is thought to be a predominant process in oxidised polypyrrole. Different response kinetics may therefore be observed for like polymers with differing anion content.

3. 3. Sensor preparation

As discussed above, conducting polymers can be prepared either chemically or electrochemically. Some while ago the gas sensing research group within the Department of Instrumentation and Analytical Science at UMIST developed a technique based on combining both approaches. Initially, a layer of conducting polymer is chemically deposited between two electrodes spaced some 100 μm apart. Of course, this is far greater than the spacing that would be amenable to electrochemical growth. Following this stage, the two electrodes are connected together and the conducting polymer is then used as one of the electrode systems within an electrochemical cell in order to grow a further layer of alternative conducting polymer upon the original chemically deposited layer. It has been found that these bilayer polymer sensors give rise to much wider choice in terms of the type of odorant molecule to which they are targeted than is possible using either of the alternative approaches on their own. Originally, these sensors were arranged in arrays of 20 elements, each one being a different material. In a commercial instrument now available, a 32-element array is employed [11].

3. 4. Interrogation

All the early work based on characterising conducting polymers employed direct current (dc) interrogation techniques. Each element is interrogated in turn by a stable current source and the voltage across the sensor element used to determine its electrical resistance. Subsequently, any relative change in resistance is recorded once exposure to an odorous compound has been arranged. A block diagram showing this arrangement is given in Figure 2. The form of response that this provides is shown in Figure 3(a). Characteristic patterns of response of this nature are associated with different odorous compounds and it is a relatively simple matter to organise pattern recognition algorithms in order to learn the various sensor responses.

Clearly, these materials also possess dielectric properties and providing the dc resistance is not too low, the dielectric property measurement may yield more information concerning the nature of the interaction between odorant molecules and the sensor material than a simple dc measurement. With this in mind, a programme of research was initiated some years ago concerned with examining the alternating current (ac) characteristics of these polymers. Typical of the form of response that can be obtained is that shown in Figure 3(b) where, for a single sensor element, we can examine its relative change in, say, electrical capacitance for a set of frequencies. Potentially, this means that we can drastically reduce the number of sensor elements and yet retain the high information content associated with large numbers of sensor elements.

3. 5. Sensor modelling

Our initial work using ac interrogation focused on simple sensor structures which were self-resonant, but at frequencies approaching 1 GHz [12]. Later, we moved to new electrode geometries which were designed to reduce the resonant frequency into the region of 63 MHz [12]. More recently, we have adopted a structure that enables us to choose resonant frequencies in the region of 1 MHz which is much more amenable to simple electronic systems for automatically and rapidly interrogating the sensors.

A lumped circuit model of such a sensor is given in Figure 4 and the model response is compared with measured data for a sensor subjected to a clean dry air stream in Figure 5. As can be seen, the model and the measured data are very closely matched and both responses exhibit characteristic resonance behaviour (where the reactance crosses the zero axis) at around 840 and 1,040 kHz. The manner in which these data were obtained experimentally is illustrated in Figure 6 which shows the automated vapour sampling system employed in conjunction with a commercially available impedance analyser (Hewlett Packard 4192A).

We have found that monitoring the change in resonant frequency is a simple and effective way of differentiating between different headspaces. For example, if we arrange to calculate the dissipation factor which is defined as the resistance of the sensor divided by its reactance, then clearly at resonance where the reactance tends to zero, the dissipation factor will rise to large values. Provided the sensor and its electrodes plus any additional electronic circuitry give rise to large electrical quality factor (Q), then these peak values of dissipation factor can be easily tracked to indicate changes in the headspace above the sensor. A simple illustration of this is shown in Figure 7 which illustrates the results of an experiment in which the odour of fresh milk was sampled at various stages over 25 hours at normal room temperature. As can be seen, a significant trend in dissipation factor emerges from this and there is clear discrimination between fresh milk and milk only three hours old. A sensory panel could not detect any difference between fresh and three-hour old milk.

Another experiment was conducted using a similar approach, but now aimed at differentiating between different types of beer. The results for this experiment are shown in Figure 8, where there is quite clear discrimination between the various beers examined. Of significance is the degree to which such measurements can be repeated and in Figure 9 we show a series of six measurements of beer samples of the same brand conducted over a period of several days which indicate that the repeatability of these measurements is extremely good.

4. Applications in the water industry

As discussed in the introduction, if gas sensors are to be useful in alerting water treatment works of potential problems in the output of potable water, we require detection limits approaching tens of ppt. No simple instrumentation method is able to achieve such a limit. To date, the best performance known to the authors results in detection limits of hundreds of ppb. However, if such an instrument is preceded by a sampling system that promotes species such as geosmin into the headspace, then this can be considerably improved. One method of achieving this is to employ an ultrasonic nebuliser which breaks up the geosmin molecular clusters in the water and promotes them into the headspace. Of course, a nebuliser also pushes a lot of water molecules into the headspace and a system such as this requires a method for removing the water molecules and leaving the geosmin behind. A technique that has proved successful is to use a Peltier based condenser above the sample and prior to the sensor array (see Figure 10)

[13]. This work, which arises out of collaboration between AromaScan plc and Mitsubishi Electric Corporation, is still in its preliminary stages and a commercial instrument will be forthcoming in the near future.

5. Conclusions

The detection of certain species which lead to unpleasant odours and tastes in water is probably best done by headspace analysis. Unfortunately, without specialised sampling procedures the detection limits required are beyond the reach of existing instrumental devices. Sampling systems capable of dealing with the problem are in the process of development and should become commercial realities in the near future. In parallel with this, there are many benefits obtained from ac interrogation of conducting polymer based sensors , not the least of which is an increase in the sensitivity of such a sensor under ac interrogation conditions. The combination of these two approaches is likely to lead to successful monitoring for species such as geosmin in water at the levels typical of the performance of the human nose.

6. Acknowledgements

This work has been supported by the UK Engineering and Physical Sciences Research Council and AromaScan plc.

REFERENCES

1. Kress-Rogers, E (Ed) Handbook of Biosensors and Electronic Noses. Medicine, Food and the Environment, CRC Press, Boca Raton, FL, 1997.
2. Persaud, K and Dodd, G Analysis of discrimination mechanisms in the mammalian olfactory system using a model nose. Nature 299, 352-355, 1982.
3. Dall'Olio, D, Dascola, G, Varacca, V and Bocchi, V Resonance paramagnetique electronique et conductivite d'un noir d'oxypyrrol electrolytique, Compt. Rend. 267c, 433-435, 1968.
4. Skotheim, T A (Ed) Handbook of Conducting Polymers, Volume 1, Marcel Dekker Inc, New York, 1986.
5. Byrd, N R and Sheratte, M B Semiconducting polymers for gas detection, NASA CR-134885, 1975.
6. Imisides, M D, John, R, Riley, P J and Wallace, G G The use of electro-polymerisation to produce new sensing surfaces: A review emphasising electro-deposition of heteroaromatic compounds, Electroanalysis 3, 879-889, 1991.
7. Persaud, K C, Bartlett, J G, and Pelosi, P Design strategies for gas and odour sensors, In: Robots and Biological Systems: Towards a new Bionics, Dario P, Sandini, G, Aebisher P, (Eds) NATO ASI Series, Springer-Verlag, Berlin, 579-602, 1993.

8. Teasdale, P R and Wallace, G G Pyrrole - A unique building block for intelligent polymer materials, Chimica Oggi 10(10), 19-24, 1992.
9. Teasdale, P R and Wallace, G G Molecular recognition using conducting polymers, Analyst 118, 329-334, 1993.
10. Slater, J M, Watt, E J, Freeman, N J, May, I P and Weir, D J Gas and vapour detection using poly(pyrrole) gas sensors, Analyst 117, 1265-1270, 1992.
11. AromaScan plc, A32S instrument, Electra House, Electra Way, Crewe, Cheshire, UK, CW1 1WZ.
12. Amrani, M E H, Persaud, K C and Payne, P A, High-frequency measurements of conducting polymers: development of a new technique for sensing volatile chemicals, Meas. Sci. Technol., 6, 1500-1507, 1995.
13. Ohta, K, Discrimination of geosmin from water using an odour sensor based on 32 conducting polymer arrays. Presented at Environmental Instrumentation Control and Automation Domestic Workshop, Japan, 26th and 27th September (in Japanese), 1996.

Figure 1 : Structure of conducting polymer

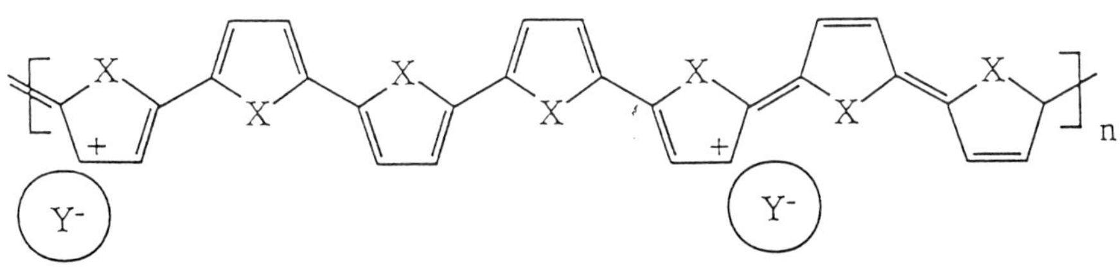

X = heteroatom (e.g. S, NH)

Y^- = counter-ion (e.g. BF_4^-, *p*-toluenesulphonate)

Figure 2 : Direct current interrogation method

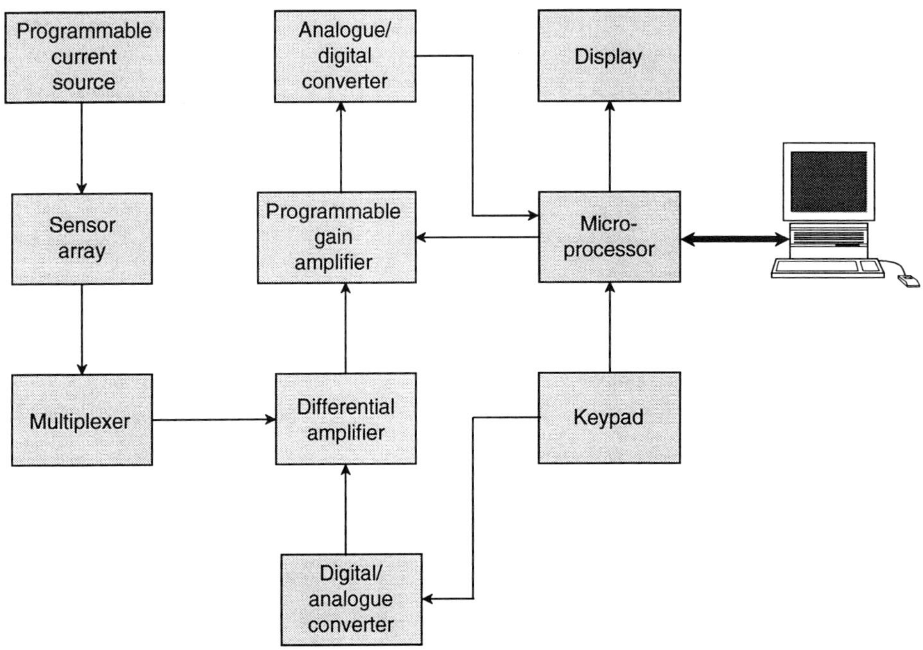

Figure 3 : Comparaison of measurements based on ac and dc interrogation

(a) Typical dc response of a 20-element sensor array

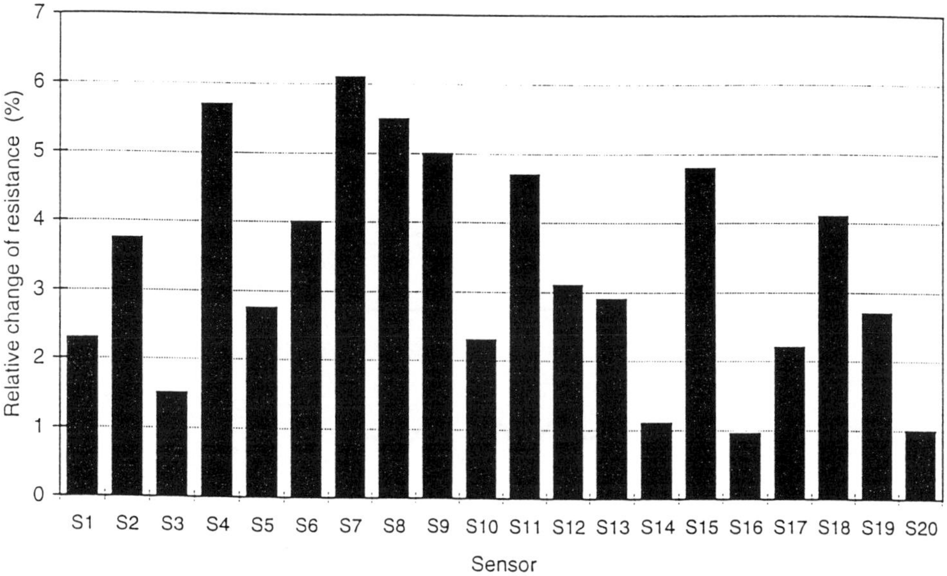

(b) Typical ac response of a single sensor

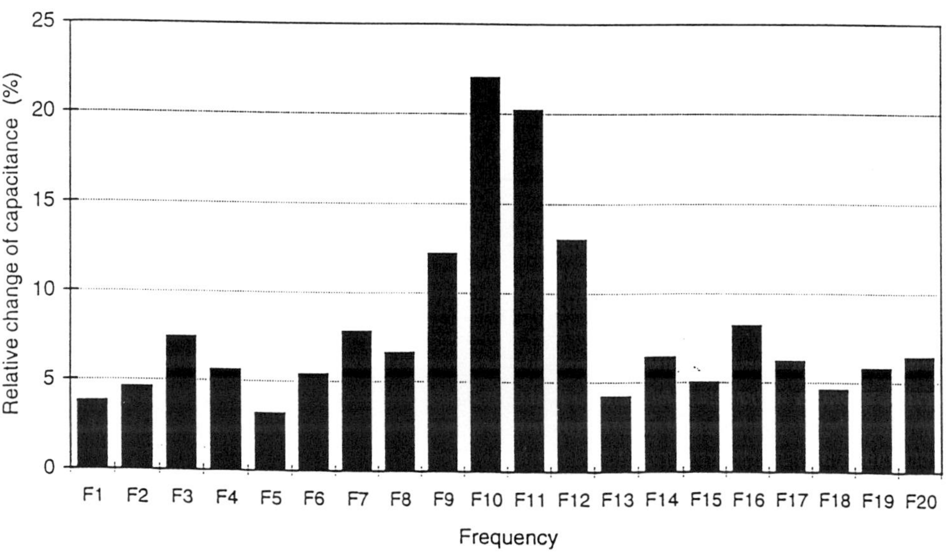

Figure 4 : Equivalent circuit for sensor showing typicaly lumped circuit values obtainde by measurements at 1 MHz

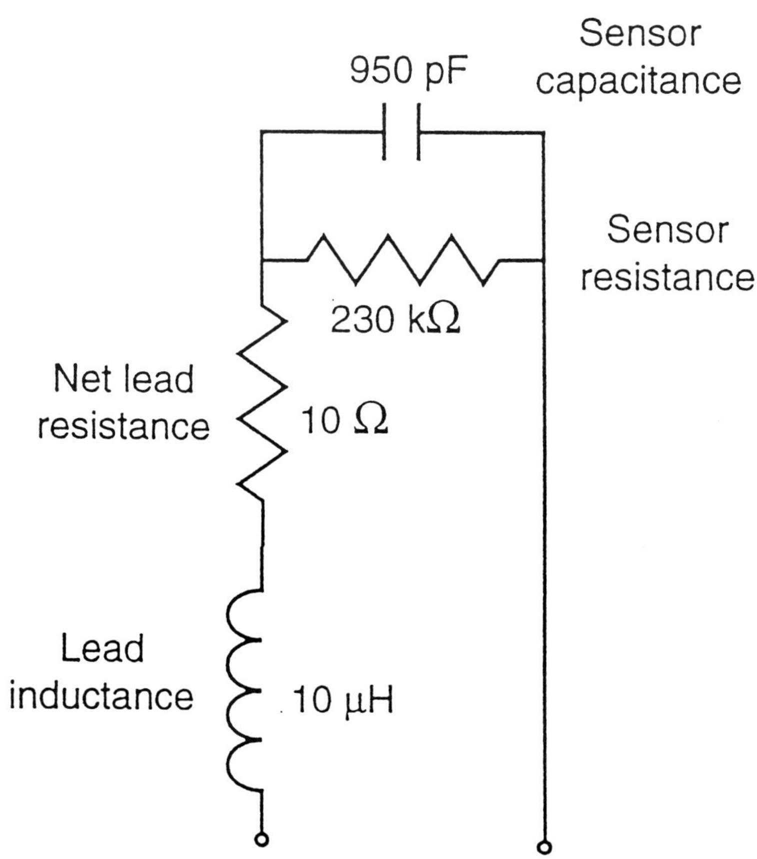

Figure 5 : Measured and calculated data

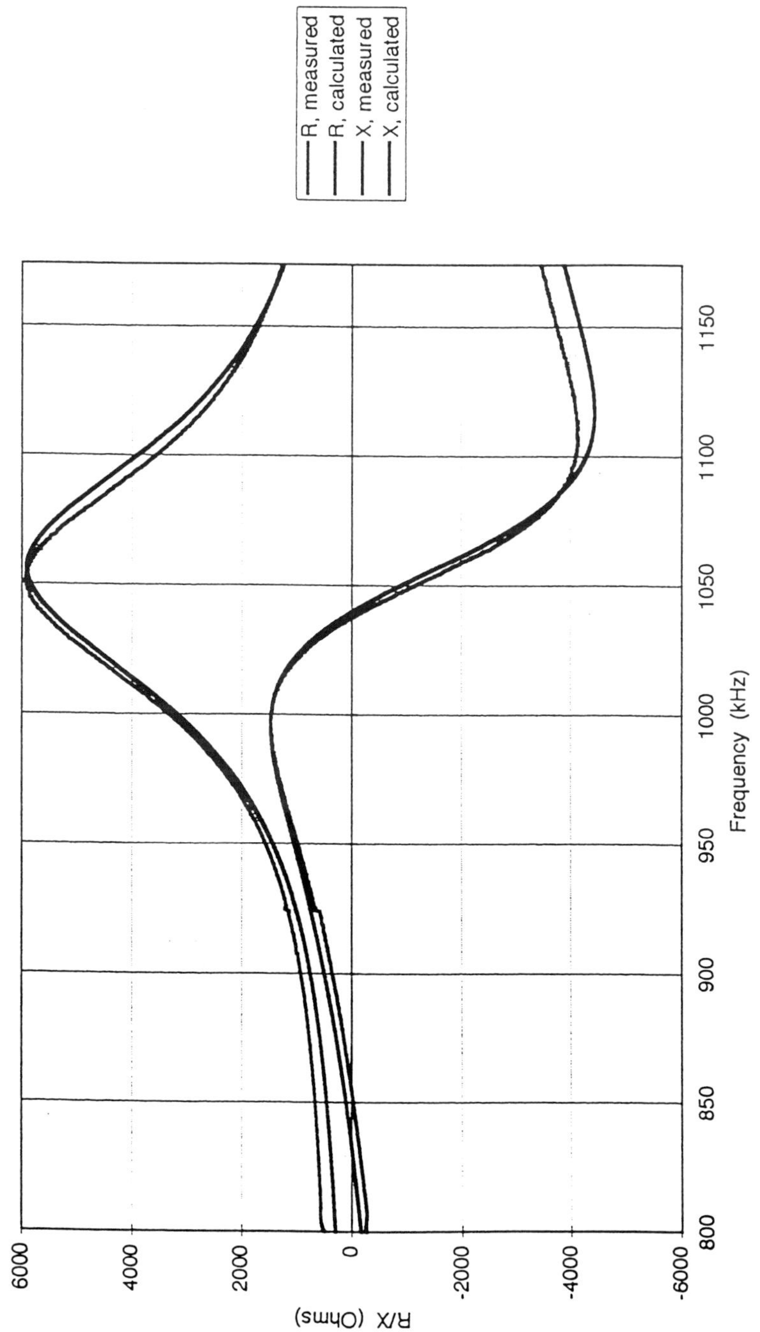

Figure 6 : Automated vapour sampling system

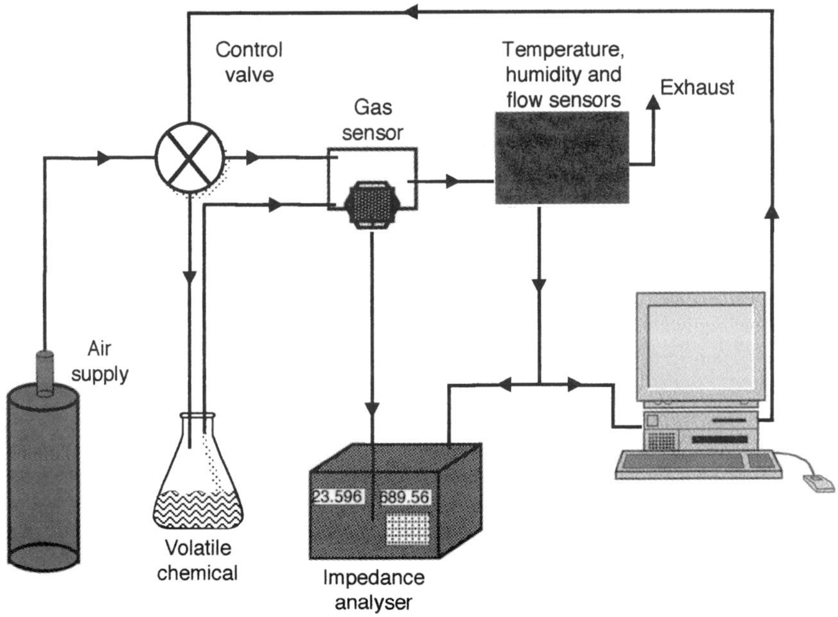

Figure 7 : Dissipation factor versus frequency for a conducting polymer gas sensor exposed to milk odour (measured over a 25 hour period at room temperature)

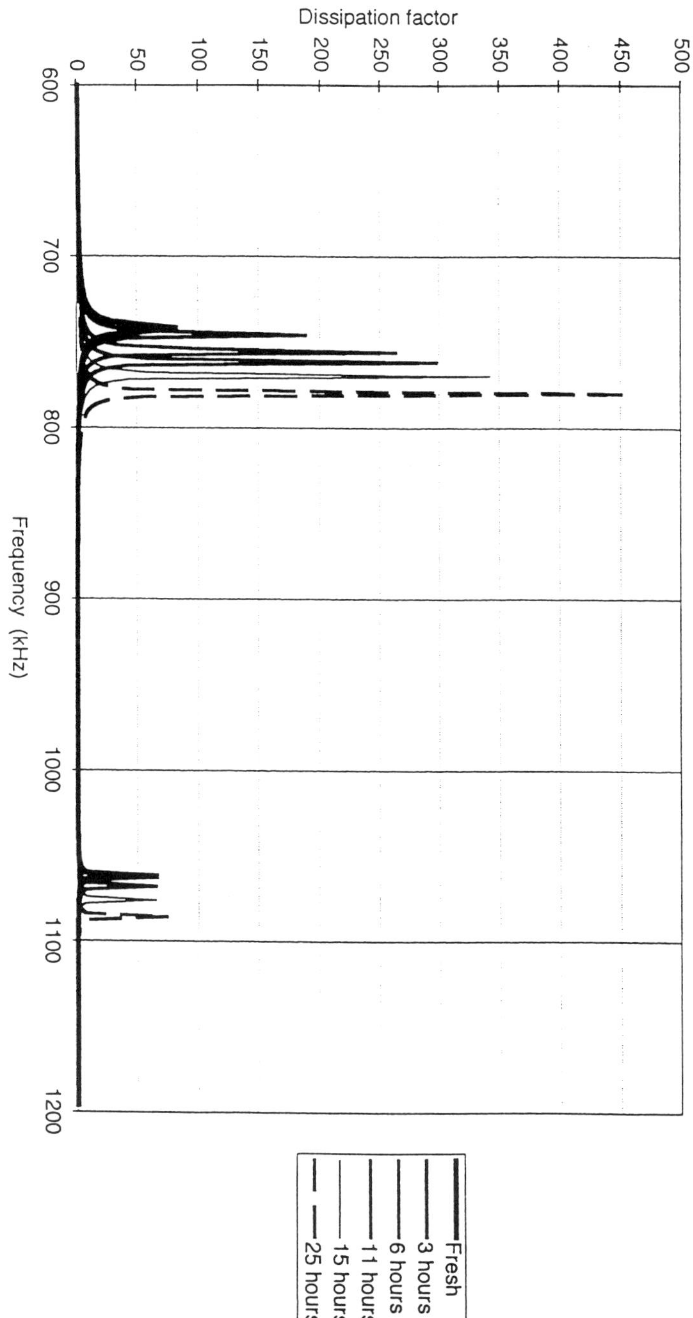

Figure 8 : Dissipation factor versus frequency for a gas sensor exposed to five brands of beer (measured at room temperature)

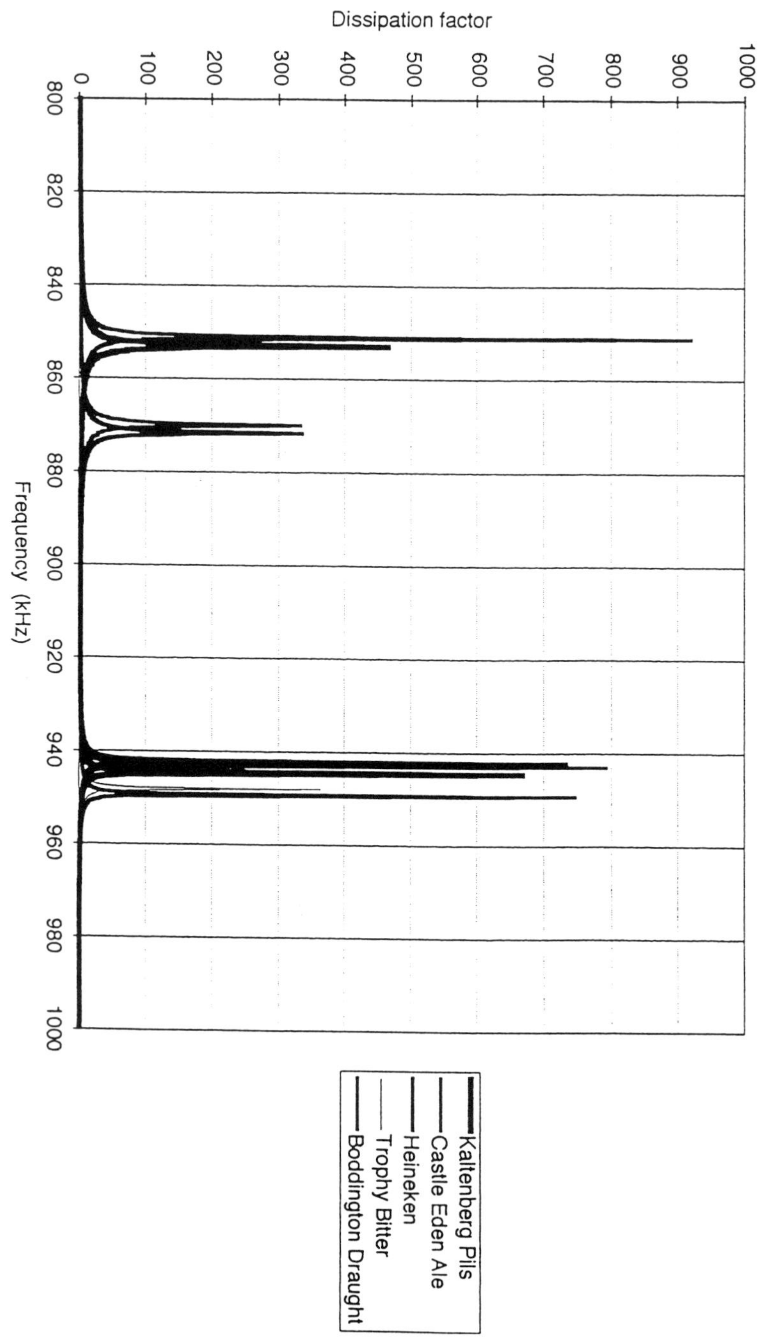

Figure 9 : Dissipation factor versus frequency for a gas sensor exposed to six differerent samples of Boddington's beer measured at room temperature over several days

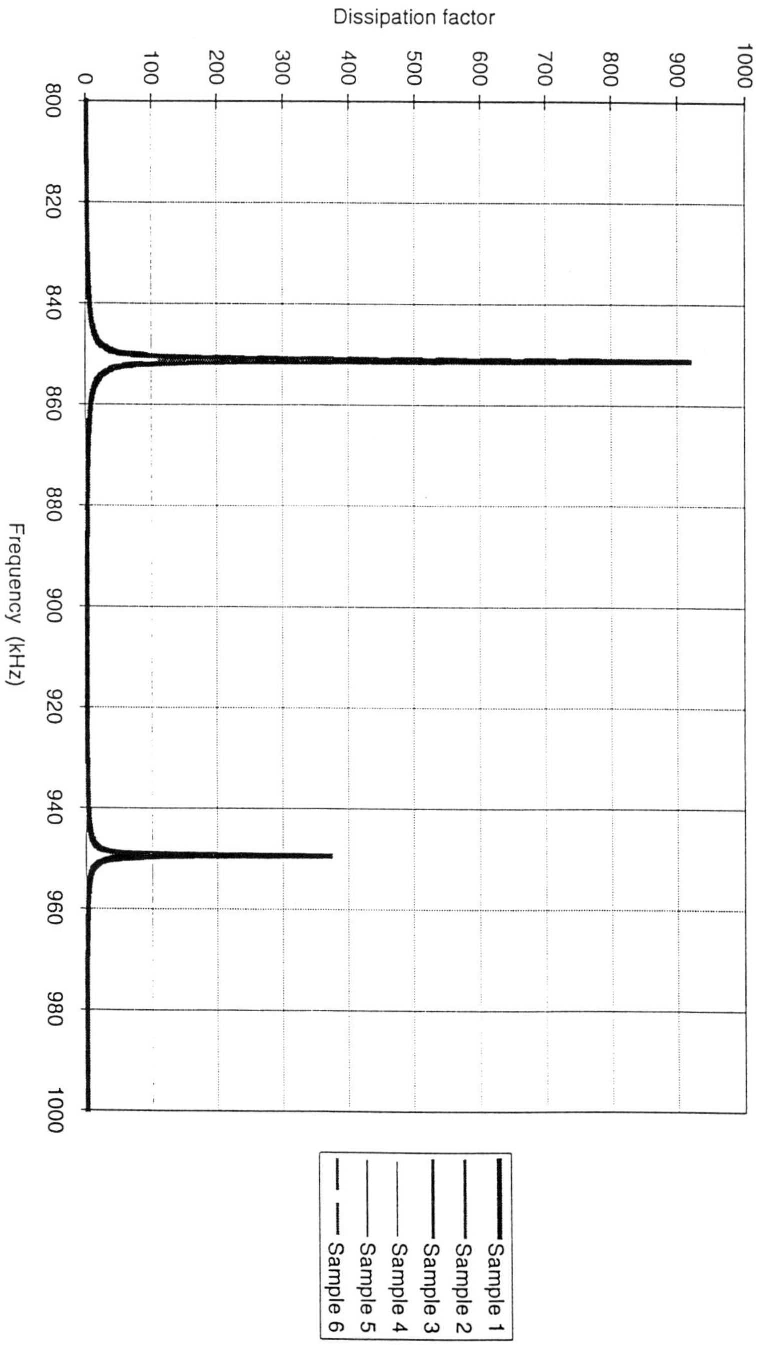

Figure 10 : Arangement for promoting a volatile chemical in a water sample to the sensor array

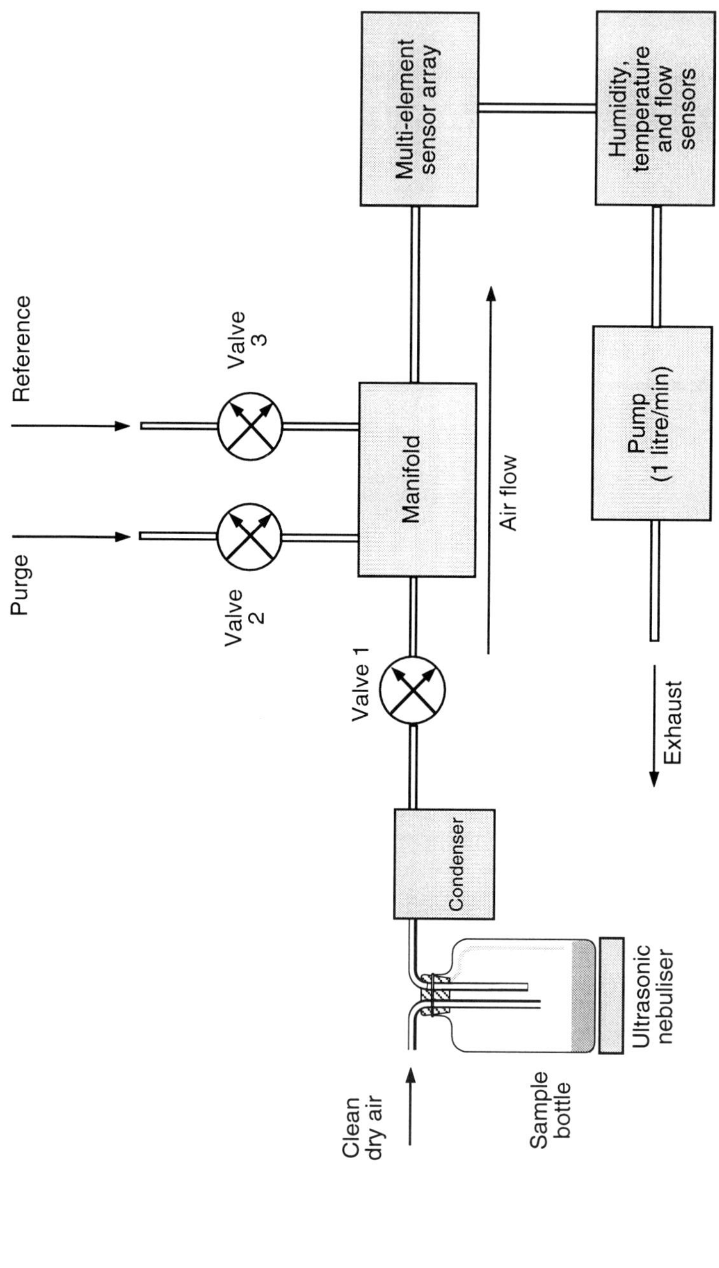

CRITICAL PH IN COOLING WATERS

By L. MARTENS, APPLITEK NV/SA, K Picquelaan 84, Deinez, Belgium

1. Introduction

In order to reduce the thermic waste from industrial cooling and process water systems, more and more industrial plants are obliged to recycle cooling water.
This recirculation concentrates considerably the levels of salts present, and some may reach the solubility limit : calcium carbonate, calcium sulfate, calcium phosphate, etc.
The soft waters, in general found in Belgium, calcium carbonate is frequently the mineral form of precipitate.
It is of course very important to have an easy and reliable method to detect in fore-hand the first signs of possible precipitation of calcium carbonate into cooling loops.

Since 1981, Laborelec (consulting agency for power stations) has developed the "critical pH-method " based on the analytical principal introduced by FEITLER in 1972.

During more than ten years this measurement has been executed in the lab by auto-matic lab titrators. Development of industrial titrators has made it possible to adapt the method for on-line measurement and monitor the cooling loops more frequently.

In the following pages we explain the principle of the method, the experimental procedure, the description and programming of the on-line analyser and the comparative results of the lab and on-line systems.

2. Measurement principle

If small quantities of concentrated base slowly are dosed to concentrated cooling water, the pH of the solution will go up first, then passes through a maximum and finally declines drastic (figure 1).

The following equations explain the phenomenon :

$OH + H \leftrightarrow HOH$

$H2CO3 \leftrightarrow H + HCO3$

$HCO3 \leftrightarrow H + CO3$

$Ca + CO3 \leftrightarrow \underline{CaCO3}$

When base is dosed in water, the OH neutralises the H + of the first reaction, what leads to a shift of the second and third reaction to the right side. The last reaction generates carbonates and frees protons. The carbonates combine with calcium to form calcium carbonate that precipitate in the fourth reaction.
The excess of protons reduce the pH of the solution.
In this phase, addition of base results in a reduction of the pH.
That maximum point is called : "the critical pH".

(see figure 1)

3. Selected material

3. 1. Measuring electrode

The determination of the critical pH can be done in two ways.

The first by measuring the turbidity of the solution, namely the latter one increases when the calcium carbonate precipitates, or secondly by measuring the evolution of the pH by potentiometric method.

The compared evolution of the pH and turbidity results in the conclusion that only micro-electrodes (ex. METROHM 6.0204.100.JE) are suitable.
The traditional pH electrodes (ex. METROHM 6.0202.100.JB) introduce measuring errors due to the thickness of the glass body of the electrode (figure 2).

(see figure 2 and 3)

3. 2. The titrator and the titration speed

All lab and industrial titration systems are actually equipped with microprocessors which control the analysis and calculate the inflection points. Most of the titrators are equipped to detect final end point or S shape curves.
The critical pH does not belong to this family.
The evolution of the pH curve is of (inverse V type).
The selected titrator depends therefore from the different titrator curves it can offer.

Whatever the choice of the instrument is, it should minimum have the following characteristics :

- monotone continuous titration
- injection of small quantities of reactive
- long time interval between two injections
- possibility to detect titration curve
- thermostatisation of the titration vessel

3. 3. Temperature set point of the sample

Significant results are only correct if the sample is kept at a temperature of two degrees above the maximum temperature of the circulation water.

T° analysis = T° max + 2°C

In order to obtain this temperature, the titration vessel is executed with a heating jacket. The jacket is fed by heated water from a thermostat bath. The precision of the regulation must be at least 0.2°C.

4. Procedure

Check up the highest temperature of the heat exchanger exit in the installation.

Adjust the temperature loop and regulate the maximum temperature of the titration vessel at 2°C above the maximal temperature of the cooling water sample.

Prepare a solution of 0.1 N base and fill the burette in order to have a correct addition.

Wait until the temperature of the sample is correct.

Start the analysis.

5. Interpretation of the results

Figure 4 represents two types of pH evolutions obtained during the measurement of the critical pH.
(see figure 4)

Generally the more the tested water sample is away from the precipitation situation, the higher is the critical pH and also the more the added volume of base will be.
On the other hand, a precipitating water type does not show any maximum.

6. On-line analyser type adi 2015

As mentioned before, the on-line analyser for detection of the critical pH should be executed with following possibilities :

- automatic sampling system (50 ml)
- dosing system (burette) for addition of precise fixed volumes of NaOH 0.1 N

- pH and temperature amplifiers
- thermo controlled analysis vessel
- possibility to program a continuous monotone titration with a fixed time between two injections of NaOH.

6. 1. Build up of the analyser Fig. 1A

The analyser is built up of 2 parts who are completely separated from each other.

6. 1. 1. Wet part of the system

Consist of :

- possibility of max. 2 burettes
- possibility of 2 sample selection valves
- possibility of 4 extra addition valves
- different type of titration vessels
- different sample systems (from 0.1-100 ml)
- max. 2 amplifier cards

6. 1. 2. Control part

Consist of :

- keyboard for programming
- power converter board => transforming current
- measurement processor board => measurement transmitters
- interfaces => control of the modules
- RS232 => data communication
- analog output => 4-20 mA

6. 2. Hardware for the critical pH application Fig. 1 B

- 1 burette of 20 ml 0.IN NaOH
- 1 macro sampling device with pipette of 50 ml
- pH amplifier + combined pH micro-electrode
- temperature amplifier + Pt 100 sensor
- jacket titration vessel 40-100 ml
- extra valve for addition of HNO3 1N (during cleaning)
- automatic stirrer (flexible shaft type)

7. Programming of the analyser

7. 1. Sequence of the basic actions Fig. 2

The complete programme for the analysis sequence is time based.
Inside the analyser is a time block that controls all activity second by second (open valve, close valve etc.).
In Fig. 2 you will find the different basic actions in one time block : RUN :

- sample selection
- sampling
- preconditioning of sample
- analysis
- calculation
- output control
- print
- rinse

7. 2. Overview of the analysis cycle for the detection of the critical pH Fig. 3

As for a complete analysis of the critical pH some steps of the former basic programme follow several times to another. Therefore we adjusted and programmed the total programme over different "RUNS".

The complete analysis cycle is build up in 4 big steps or runs.

More in detail we can summarize the programme (see Fig. 4).

RUN 1 : Sampling

RUN 1 consists of :

- sample selection
- rinsing of the macro pipette with fresh sample
- filling of the burette with NaOH 0.IN
- empty + 3 x rinse of the analysis vessel
- ml fresh sample inject in the analysis vessel

RUN 2 : Preconditioning of the water sample

The preconditioning of the water sample : in this case consist of heating up the sample with 2°C higher than the max. temp. of the circulation water :
T° analysis = T° max. + 2°C

RUN 2 consists of :

- measuring of the temperature of the sample (Pt 100)
- "if statement" : if the temperature of the sample is equal to the requested
- temperature, then we go th RUN 3.
- "if statement" : if the temperature of the sample is not correct, then RUN 2 is start up again.
- measuring the initial pH of the sample

RUN 3 : Analysis

In this run the final analysis of the critical pH is done (monotone titration).
A fixed volume of 0.1 ml NaOH 0.IN is added to the sample and after a waiting time of one minute the pH is measured.
This result is memorized in the analyser and again 0.1 ml NaOH is added to the sample. After one minute the pH is measured again.
By taking the difference in pH of 2 successive measurements the "delta pH" (pH) is calculated.
As long as this pH > O the system goes on with dosing NaOH (always in steps of 0.1 ml).
If the pH reaches < 0 then this means that the critical pH is achieved and we can go to run 4.

RUN 3 consists of :

- dosing of 0.1 ml NaOH 0.1N (by burette)
- minute waiting time
- measuring the pH
- calculate the pH
- "if statement" : if the pH > 0 go to RUN 3
- "if statement" : if the pH > 0 go to RUN 4

RUN 4 : Result processing

This run consists of the processing of the different results and the print out of the requested information.

RUN 4 consists of :

- addition of 10 ml NH03 1N to the sample (by valve) to avoid problems with the precipitated carbonate in the next analysis.
- printing out of the results : initial pH, critical pH and dosed NaOH to attempt the critical pH.

Remark :

In Fig. 4 is a schematic print out of the different actions within the 4 runs and this action by action. In the final programming of the analyser it is of course possible to programme several actions at the same time.

The software of the analyser is build up in programme lines. That means that for each valve (max. 3 + max. 4 extern valves) and for each burette (max.2) one line is available, further there is a programme line available for the drain, sample loop, the analysis, the calculations, the print out and the analog outputs.

The total software package of the ADI 2015 is realised in 19 programmable lines. The programme is composed over 2 different programme sheets (see Fig. 5) : one with the different time blocks (time programming sheet 1) and one with further details for the different actions programmed within this specific time blocs (time programming sheet 2).

For a valve this means that we only on "sheet 1" find a line and nothing on "sheet 2" for a valve it is only possible to open or to close.

For a burette we can find lines back on both sheets. On the first sheet we have to fill in which tune block we want to use a burette and on the second sheet we are programming the action we want (filling, dosing or cycling ...).

For the analysis itself that means that on sheet 2 the different titration parameters have to be indicated.
Once a complete programming of the ADI 2015 is done, we can test the application.

8. Results of testing slide 2a, 2b

8. 1. Test of reproducibility using

- 0.05 N NaOH
- 0.1 N NaOH

8. 2. Parallel test of lab titrator and on-line titrator with conc. NaOH 0.05 N

Volumes for both instruments are not very reproducible.
pH values are acceptable. The choice is 0.1 NaOH

8. 3. Results water sample SERAING

- Slide 3A at 40°C
- Slide B 50°C

8. 4. Results water sample "Pont Brulée" without metagon at 45°C

8. 5. Results "Pont Brulée" without metagon

Conclusion : Both systems work correct.
The on-line system works sometimes better reason :
On the ADI 2015 is an automatic system.

Figure 1 : Evolution of the pH during the determination of the critical pH

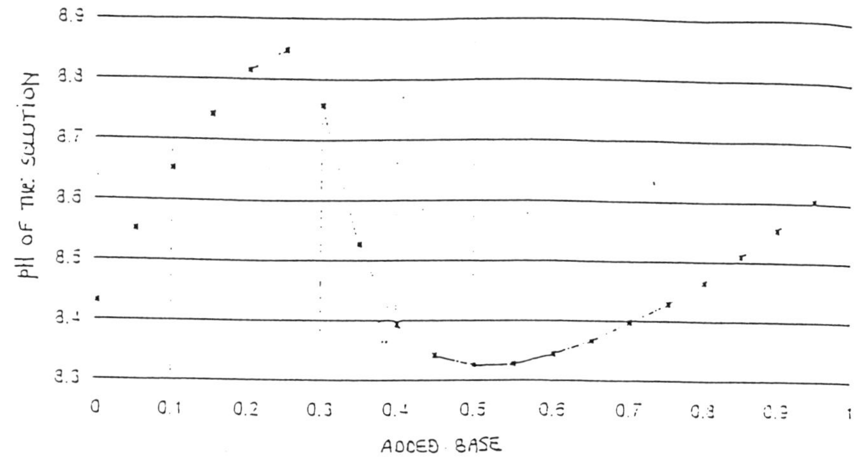

Figure 2 : Evolution of the measured pH in function of the used type of electrode

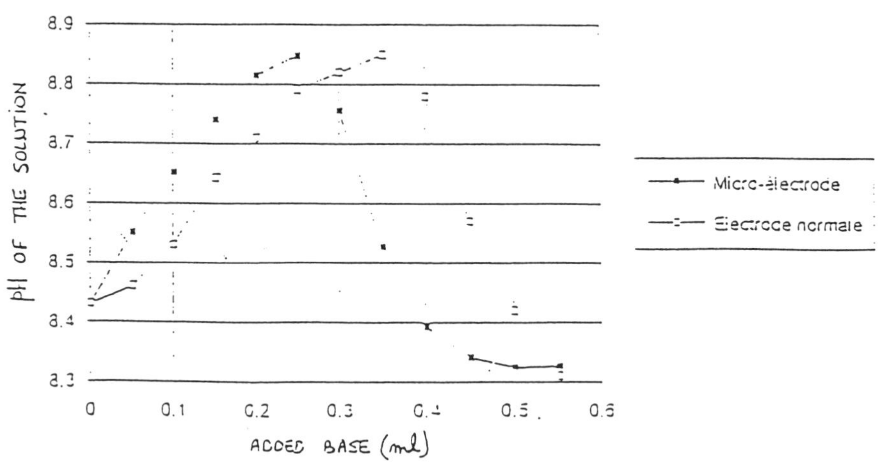

Figure 3 : Evolution of the turbidity during the determination of the critical pH

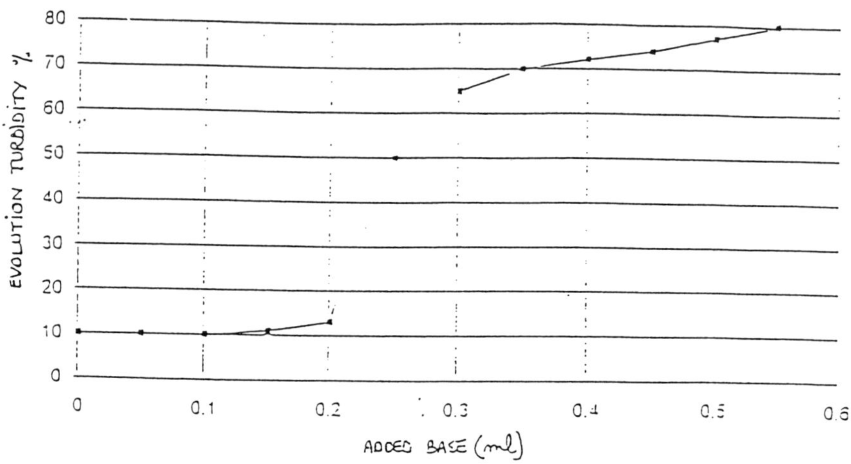

Figure 4 : Evolution of the pH in different situations of the water

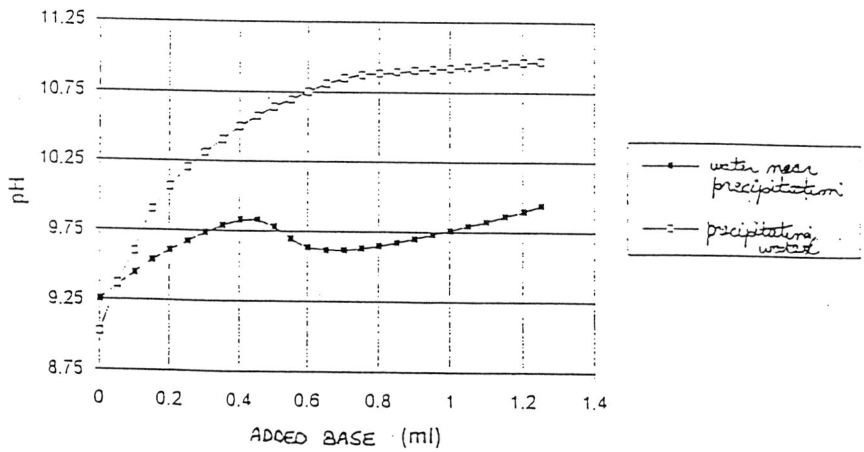

ADVANCED TECHNOLOGIES FOR NEW PARAMETERS AND MEASUREMENT CONCEPTS

M. JAUZEIN, Director of the GIP Stelor and Research Director of IRH Environnement, 11 bis, rue Gabriel Péri, 54 500 Vandoeuvre, France

1. Introduction

The development of advanced technologies for data treatment in the field of computer science and numerical tools gives a sound basis for information structuring, analysing and drawing and for the understanding, modelling and validation of parameters relationships.

The monitoring of water quality can be performed through a large set of classical parameters, innovative highly specific measurements, and sensitive sensor signals. But then the number of individual data and the interpretation of less specific informations become new limiting factors. To overcome these limiting factors, advanced technologies for data treatment can be optimised for the derivation of virtual or integrated parameters which are the keys of observed time or spatial behaviours.

The phenomenological approach is generally used for identifying relationships between variables on the basis of deterministic models describing elementary phenomena steering studied behaviours (thermodynamical laws, mass balance constraints, kinetic laws, ...). In the case of physico-chemical parameters describing the quality of water, these relationships are generally complexes and often non linear including additional parameters that are not known only on a statistical basis or with poor confidence. This general approach gives many rules for understanding multivariate behaviour of real systems in surface water and groundwater monitoring, pollutant transport in natural waters, drinking water or sewage collection networks, and water treatment processes. Nevertheless, the deterministic approach is limited by the lack of confidence for some additional parameters due to insufficient data basis or uncontrolled parameter fitting.

The " black box " approach is an alternative to this phenomenological approach. Without a clear knowledge of the phenomena steering the multivariate behaviour of real systems, it consists in deriving key parameters from a large set of data characterising a studied object. The proposed data treatment strategy has the aim of reducing the set of data to an optimised set of integrated or virtual parameters describing the main shapes of the multivariate systems. Multivariate statistical analysis is one of the more classic approach that has been applied in this context. This first pathway for solving the problem is limited by the fact that only linear correlations are analysed. The non linear relationships need to state preliminary data transformations using empirical or phenomenological non linear laws. Other pathways are now open using, for example, virtual connections through neural networks or fuzzy logic rules.

This paper has not the aim of giving a theoretical exhaustive view of those advanced numerical techniques but to illustrate their potential use in the field of water quality monitoring.

2. State of the art

The multivariate linear statistic analysis is one of the most classic way to reduce information packages to principal factors or components. Thus, it is not necessary to remind those techniques, but only to outline the limits of this approach for taking into account the existence of non linear relationships between variables.

2. 1. Neural networks tools [1 to 11 and 24]

The human brain is able to memorise informations, to classify them, to relate and identify them. Through the observation of the biological mechanisms governing brain functions, several fundamental principles has been identified which allow to reinforce existing techniques and develop innovative data treatment methods. These tools are based on a neuromimetic or connectionist approach. As a mater of fact, neural networks try to mimic neural structures through connection networking. The high number of connections, the non linearity, the relationships between input and output parameters are the main characteristics of those numerical tools. The connectionist term is more appropriate for outlining the distance between present neural networks and the biological model.

In the present state of the art, connectionist methods are not universal solutions, even they can be applied successfully in many practical applications. In the future, the combination of neural networks with expert systems, classical algorithms and complementary tools will generate the more interesting data treatment technologies.

The basic element of a neural network mimic the cell of a brain system. The characteristic functions of these cells are the reception of signals from adjacent cells, the integration of these signals, the production, transport and delivery of a new signal to other cells. Without giving more details on the biological model, it is important to remind that the cell integrates continuously the input signals through synapses connections. To mimic these basic characteristics, the main idea is to use basic elements connected in networks through Artificial Neural Networks : a graph where nodes are elementary cells and arrow are connections. Cells can be entirely networked or structured in several layers.

The layered neural networks can be compared to statistical multivariate analysis. Specific neural network structures are exactly performing Principal Component Analysis or Discriminent Factorial Analysis. The main advantage of neural networking is to allow the integration of non linearity in the optimisation process through the activation function transforming the input function in a different output function. If preliminary non linear transformations can be applied before statistical multivariate analysis the problem of selecting them is unsolved. For neural networks, the selection of simple non linear activation functions makes the design of the data treatment easier.

But, the classic multivariate analysis gives more interesting results in terms of confidence analysis and correlation indices compared to present neural network tools. Another limiting factor is the selection of the optimal structure for the neural network. A strict optimisation strategy is generally too time consuming and a empirical approach is applied.

These ANN tools are directly applied to data reduction problems. In the case of time dependent series, it is possible to use as input data the multivariate signals for different past time data series to estimate an output data set at the present state. Similarly for space dependent data, there is potential for developing interpolation strategies based on the use of nearest multivariate data set to estimate an output data set on a specific location.

2. 2. Fuzzy logic tools [12 to 23 and 24]

The fuzzy logic approach, introduced by L.A. Zadeh in 1965, allows to treat informations easily with poorly defined concepts. This approach mimic human cognition processes. In addition, the concept of fuzzy object allows to define the link between this object and a well defined category of objects. For example, the limit between cold and hot temperatures refers to individual sensations which can differ between persons. The relationship between the words cold or hot and the personal feeling of each individual from a population includes a significant level of imprecision. This fuzzy limit between the category of situations cold and hot needs to be quantified. The fuzzy logic approach consist in establishing rules for linking the " hot " domain to numerical rules such as " temperature from 40 to 100°C ". In the case of misunderstood phenomena, this approach can be interesting compared to the deterministic one. The fuzzy definitions is a basic characteristic of human cognition. In many situations, fuzzy informations, fuzzy rules of interpretation and decision are used by human beings. Thus, the fuzzy logic tools allows to combine fuzzy data and precise numerical data in the process of data treatment and related decision making. Compared to conventional numerical limits, it also allows to take into account their imprecision through formalised precise rules. As a consequence the fuzzy logic concept is based on a precise numerical methodology to describe fuzzy systems.

A fuzzy set of objects is a class of object characterised by a continuous inclusion degree law ranged between 0 and 1. Compared to classic sets of objects where the inclusion law is binary (0 for excluded objects and 1 for included objects), this concept is more flexible. Fuzzy rules has been introduced by Mandani in 1974 for monitoring real systems. They are based on " if ... then ... " logic algorithms and are potential formalisms for developing expert systems on automated cognition processes. The main difficulty of this approach is to define the number of necessary rules which can be wording rules or functional rules (numerical law s).

The use of fuzzy logic for describing multivariate systems has appeared more recently It's a complementary tool for studying non linear or complex multivariate situations. These numerical tools are precision and complexity limited but their optimal design is very efficient for qualitative analysis of real systems. As neural network approach, it allows to derive output key variables from complex input data sets without a clear phenomenological description. Recently, a theoretical study of fuzzy logic systems has shown that they are universal estimators or mono-variate functions and efficient estimators of multivariate functions. Some publications are also dedicated to optimised decomposition of linear system in fuzzy sub-systems. Finally, the design of non linear relationships is easier through fuzzy approaches. In addition, the fuzzy approach is available for interpolation procedures in the case of time dependent series and thus potentially for space dependent variables.

To perform the initial fuzzy analysis of a system, it is necessary to transform the variables in the form of fuzzy ones : this stage is called the fuzzyfication of the problem. Then the variables are treated with fuzzy logic rules and then obtained fuzzy results are transformed in the form of numerical data results : this last stage is called the defuzzyfication.

3. Monitoring of a water treatment plant with time dependent series

3. 1. The water treatment process

The water treatment process which has been studied is a classic multistage activated sludge process including oxygen injection systems for the activation of biological degradation phenomena [25]. It includes three identical bioreactors used for the continuous flow treatment of waste water fluxes. Each bioreactors can be steered for activating specific degradation processes depending on the quality of the waste water which is time dependent on an unpredictable way. Monitoring systems are installed for the following of pollution parameters and to control the treatment process.

3. 2. The data available

A specific set of parameters has been selected for the case study. It includes measurements of specific pollution parameters, biomass monitoring parameters, physical parameters (temperature and flow rates) and process control parameters (oxygen and biomass increase). Due to confidential reasons, the case study is presented with letters for identifying parameters (A to K for input parameters and L to N for output parameters).

The main set of variables are characterised by a cyclic evolution. The variables A to G show similar time behaviours but the sampling period differs for each parameter.

3. 3. Comparison of three advanced data treatment strategies for monitoring the water quality

Three methods have been used for time dependent series which has been preliminary normalised on the basis of the average, the maximum and minimum values :

- the auto-regressive approach using input data (AR) or a moving average approach including output data treatment (ARMA),
- the neural networks approach,
- the fuzzy logic approach.

The auto-regressive approaches are based on multi-linear regression tools applied to finite time series of input data (AR) or finite time series of input and output data (ARMA). The implementation of this approach needs to define a learning data base, a validation data base, a design of individual model structures, the fitting of parameters on the learning database for each model and the selection of the best model using a single criteria applied to the use of the model on the validation database. For this type of numerical tools, the ARMA approach is more efficient then the simple AR procedure for the same number of additional parameters.

The neural network approach has been applied through the test of three layers systems : one input layer, one intermediate layer and one output layer. The input set of data can include input and output past time series. The selection of input data sets has been performed on the basis of previous results obtained with the ARMA approach. The main results obtained is that, for the same number of parameters the results obtained with neural networks are better than using classic ARMA methods. This seems to be related to the easiest description of non-linear behaviours with neural networks than with multi-linear regression analysis.

The test of fuzzy logic approaches has been performed on only one input data and one output data. The main results obtained are a good estimation though learning runs, a high number of rules (about two thousand) which needs a prioritisation strategy, and a failure in the validation process.

This comparison of three type of advanced techniques for time series data treatment on a real system has given interesting preliminary results. Using present and past input data and sometimes past output data, these methods can be used for the prediction of key parameters for the monitoring and control of waste water treatment processes. The preliminary results obtained show that :

the inclusion of past output data jointly with present and past input data gives better results for the same number of input parameters.
the Auto-Regressive Moving Average and the Artificial Neural Network approaches are efficient compared to the fuzzy logic approach.
the Artificial Neural Network seems to be more adapted to this specific application on time data series.

A sequential use of ARMA and ANN is adapted to the multivariate analysis of time dependent behaviour. A more complete use of present input and past input and output data with this sequential approach gives a sound basis for future developments. Additionally, the development of moving learning procedures can be studied for on line correction-prediction strategies adapted to process phenomena evolution. Concerning fuzzy logic approaches, the fuzzyfication process has to be optimised to generate a limited number of key rules which will be easier to interpret and manage for deriving interesting data treatment alternatives.

3. 4. Test of a multi-model approach using fuzzy logic

A multi-model structure has been develop on the basis of a fuzzy logic approach to study the multivariate behaviour of the tested system. A specific formalism generalised from " Hammerstein " has been tested and a specific parameter fitting techniques has been developed. This formalism includes a finite number of rules based on non-linear static models coupled with a single linear discrete transfer function for the dynamic part of the relationship.

The combination of simple models which are efficient within a small range of process running conditions is designed through a specific aggregation rule able to give an algebraic relationship between input data to each output parameters on the whole range of process running conditions.

The fuzzification of each parameter is performed as a preliminary data treatment and consists in a partitioning of numerical data. The establishment of fuzzy rules is performed through a multi-model structuring of the relationship between input parameters and each output parameters. Then, the defuzzification is processed through classic numerical procedures giving a result for each elementary model rules. Then an aggregation law is used to derive the global estimation of each output parameters.

The developed identification procedure for parameters is based on conventional non-linear equation iterative solving (Gauss-Newton or Levenberg-Marquardt methods) applied to the formalism generalised from " Hammerstein ".

For the application of this approach to the presented case study, normalised data has been employed. Then some exotic values has been eliminated and a moving average filter has been applied to the original data set. The selected output data for illustrating the method is the UV absorption parameter.
(see figure 1 and figure 2)

The first step is performed through the optimisation of monovariate analysis between each input parameter and the studied output parameter. A prioritisation of obtained correlations is obtained and allows to perform a step by step identification of multivariate descriptions. In comparison with classic multivariate linear statistic analysis, this techniques allows to classify the interest of each variables as principal components and select the number of parameters on the basis of the increased correlation coefficient. From a correlation coefficient of about 0.88 obtained with a monovariate model, it increases up to 0.98 using all the available information. But three variables are sufficient to obtain a correlation coefficient close to 0.94 (see figures n° 1 and 2). The three variables multi-model has been analysed in terms of structure of the " Hammerstein " generalised model and sensitivity to each parameter.

The study of the dynamic part and its influence on the correlation coefficient indicates that the last point in the time series is optimal giving the maximal correlation and that the optimal order for this dynamic part of the model is 1. Then the number of rules has been studied giving 3 as an optimal number just through graphical determinations.

This learning procedure applied to one monitoring campaign has been validated on the basis of two other campaigns. For one of these campaigns, the results are promising and for the other, characterised by higher perturbations, the results are less convincing due to a transient drastic change in the quality of waste waters.

This study outlines the efficiency of the multi-model fuzzy logic approach for the estimation of multivariate and non linear behaviour of real systems. Some enhancements are needed for the optimisation of the multi-model structure. Lastly, one can note that the UV absorption parameter is the expression of the residual organic matter content in the effluent which can be related to the chemical oxygen demand (COD) and used for a feed-back control of the process.

Figure 1 : Simulation results for B (UV) obtained with 1 input parameter (G)

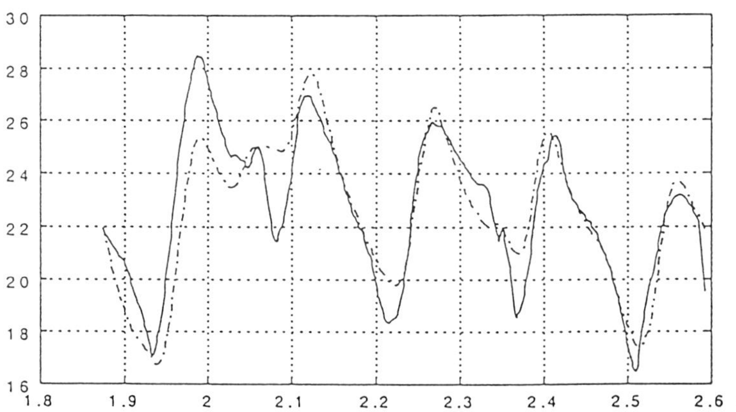

Figure 2 : Simulation results for B(UV) obtained with 3 input parameters (G, F and B)

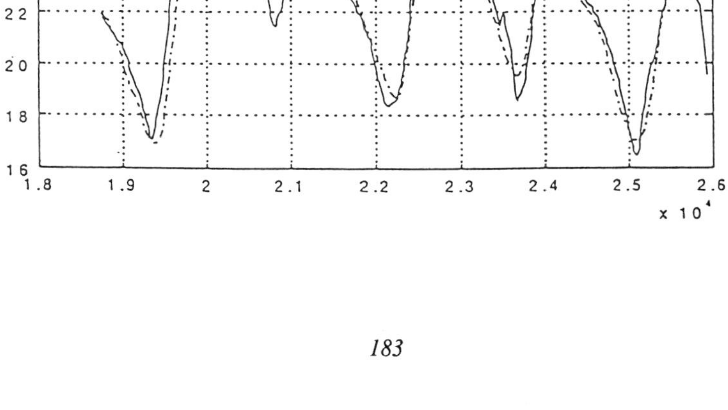

4. Monitoring of groundwater quality with space dependent series

4. 1. The groundwater pollution source

Waste deposits are one of the pollution sources generating an acute or chronic degradation of water resource quality. For the assessment of contaminated sites, it is necessary to obtain maps giving a clear idea of the space distribution of contaminants. The selected case study is a waste deposits containing a mixture of municipal and industrial wastes [26]. A large range of organic solvents has been detected in the leachates of this site. These solvents are mainly mono-aromatic compounds like toluene (BTEX compounds) and halogenated aliphatic compounds (HOC compounds).

4. 2. The data available

In the case of the presence of volatile organic compounds (VOCs), it is possible to use global sensors delivering data on the presence and concentration of contaminants in the soil and subsoil gas phase. The main advantage of this sensors is the possibility to design field screening campaigns based on a large number of localised measurements obtained on the basis of a regular grid. The main limit of these sensors is the delivery of a global signal from each sensor, which is generally highly sensitive but only poorly selective. To overcome this problem, the use of several sensors at the same time can be proposed for delivering a multivariate signal which can exhibit a better selectivity. Another limiting factor for the use of this screening techniques is the indirect relationship with real groundwater contamination. As a matter of fact, even the partitioning of VOCs between the gas and the aqueous phase is linear, the relationship between the measured signal and the groundwater contamination is not necessarily linear. In addition, the presence of VOC mixtures with a large range of physico-chemical properties generates non-linear dependencies for global measurement systems.

To obtain significant maps of the contamination of groundwater by organic solvents, the following methodology has been applied : combined measurements with three VOCs global sensors on about 80 localised points, space dependent data treatment of results, selection of about 25 contrasted points, sampling and analysis of groundwater at this selected points, correlation analysis with previous results for estimating the relationship between VOCs mapping parameters (input data) and groundwater regulation parameters (output data).

The VOC sensors are a photo-ionisation detector, a semi-conductor sensor and a lipidic membrane piezoelectric sensor. The measurements have been performed using the installation of fixed perforated tubes down to a 3 metres depth. The gas phase is equilibrated with the subsoil gas phase during more than one day in the closed tubes. Then the gas phase of the tubes are characterised with the three sensors. Concerning aqueous phase sampling and analysis, the same tubes has been used for leachate sampling and the analysis of a set of priority pollutants have been performed through gas chromatography and specific detection of individual VOC.

4. 3. The methodology developed for the space dependent data treatment

Due to the fact that each detector gives a complementary response which is partially redundant with the others, a classical multivariate statistical analysis can give a better idea of the data redundancy determining principal components and their contribution to the global variance of the system. Consequently, this approach has been followed to identify new virtual parameters using a Principal Component Analysis (PCA) on the basis of normalised data sets. The result of this data treatment is the prioritisation of three virtual parameters with the help of variance explanation indices. Compared to initial parameters, these virtual parameters are characterised by their orthogonal behaviour (and gaussian behaviour if possible) for the studied data set.

Then a set of maps can be obtained for each virtual parameters to visualise the main characteristics of spatial heterogeneities (see figure 3). The proposed method is based on kriging after a variographic analysis of space dependent heterogeneities. The same type of model variograms (spherical) with different additional parameters (nugget effect, scale and maximal variance) has been derived from experimental data. Then these model variograms are used for interpolating data and delivering kriging maps of virtual parameters distribution.

4. 4. Correlation study with conventional groundwater analysis

The combination of field VOCs measurements (virtual parameters) and laboratory results on a single set of 25 localised points has been studied with a principal component analysis. Consequently, specific individual or groups of analytical variables can be correlated with the identified virtual parameters. Using the first three principal components and the three virtual parameters, it has been possible to determine optimal multi-linear models to transform virtual parameters (input data) into estimators of specific or groups of analytical variables (sum parameters). A parallel combination of the characteristic weight of the virtual parameters in the PCA allows to determine a confidence criteria for each multi-linear model. For example, a specific individual components highly correlated to a virtual parameter with a high weight can be estimated with a high confidence criteria (close to 1 as for benzene in figure 4). On the contrary, if the parameter to be estimated is only correlated with a virtual parameter characterised by a small weight, the confidence level will be too small (close to 0).

This specific study is an illustration of the combination of multivariate and space dependent statistical analysis in the field of water quality monitoring. The tools that has been tested are normally devoted to gaussian variables and linear relationships between variables. Thus, the development of alternative approaches based on neural networks or fuzzy logic for the studying of space dependent variables could be promoted. In addition, the proposed approach outlines the interest to develop multi-variate sensor systems based on spectral signals or multi-sensors arrays and the crucial need of a validation stage for deriving confidence criteria of produced maps.

Figure 3 : Maps obtained using the first principal component as virtual parameter

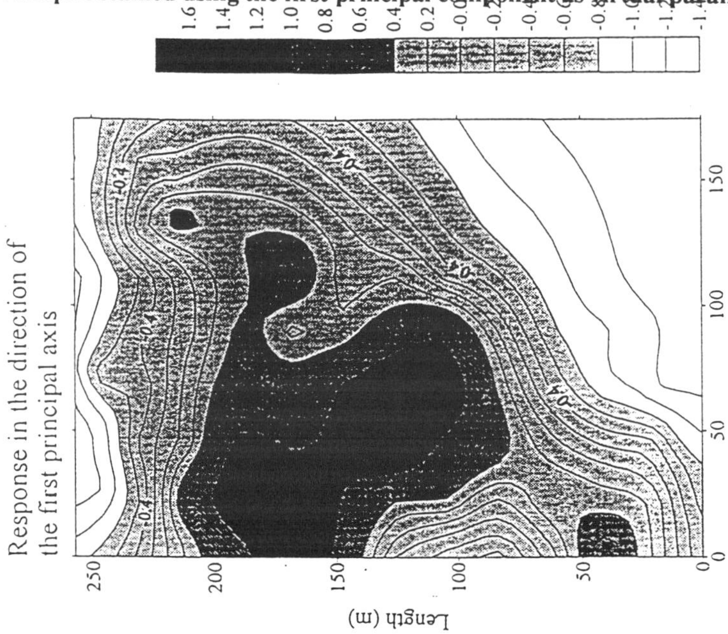

Figure 4 : Comparaison of a benzene concentration estimation and measured concentrations in leachates with a 0.90 confidence indice.

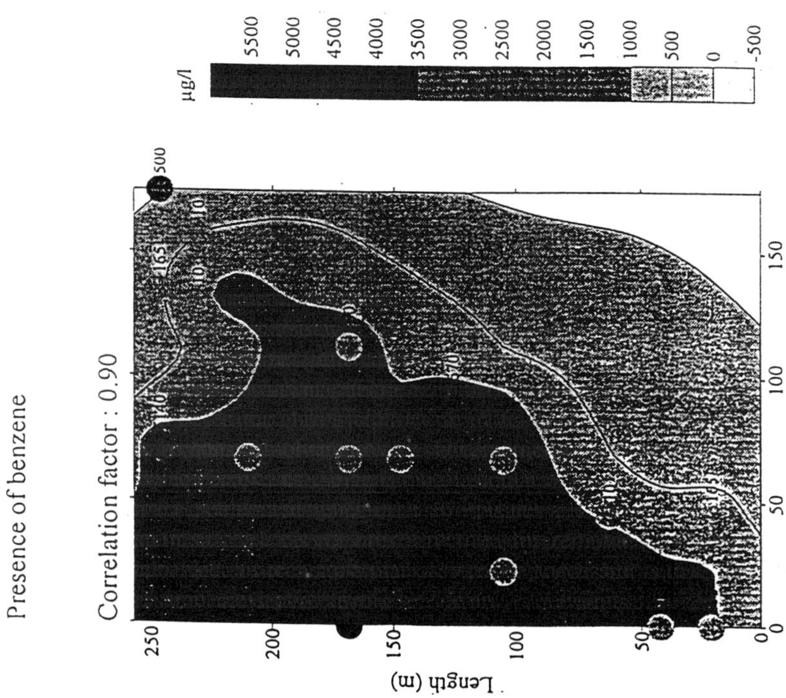

5. Conclusions And Perspectives

Three main types of chemometric multivariate techniques has been presented :

classic linear statistical analysis including multivariate statistics, multi-variate regression approaches for time dependent variables and geostatistical approaches for space dependent variables,
neural network approaches,
fuzzy logic approaches,

These advanced technologies for data treatment has been illustrated through case studies concerning a waste water treatment process and a contaminated groundwater plume. The main conclusion of this scientific and technical note is that the presented tools are now available for specific demonstration in the field of water quality monitoring either for time dependent or space dependent parameters. The range of application fields is large and the parallel development of sensor arrays of spectral signal sensors will increase the interest of these efficient numerical tools. To derive virtual or integrated parameters directly available for decision making from large and complex data sets, the optimal combination of phenomenological, statistical, neural and fuzzy approaches seems very attractive for future research and development in the field of measurement and testing or in the field of advanced technologies.

6. Acknowledgements

I acknowledge F.D. RAVOIRE, J.Y. CATROS, V.QUELIN and E. RANNOU from THOMSON-CSF, Services Industrie, J. RAGOT and C. LEVERINI from the CRAN-CNRS of Nancy (Centre de Recherche en Automatique de Nancy) and F. COLIN and G. GRAPIN from IRH Environnement in Nancy for their important contribution to this paper as partners in a joint research programme financed by the Ministry of the Environment (DGAD/SRAE/94059) on “ black box ” modelling approaches. I also acknowledge P. ROCKLIN from the NANCIE (International Centre for Water of Nancy) for its PhD contribution in a specific field application financed by the Ademe (National Agency for the Environment and Energy Management). Finally, I want to outline that a part of this work has been founded by the “ Environment and Climate ” european research programme (DGXII) in the field of “ technologies for the environment protection ” for the measurement and modelling of the VOCs mobility in soil and groundwater systems.

REFERENCES

1. E. Davalo and P. Naim, Editions Eyrolles, deuxième Edition, Des réseaux de neurones (1990).

2. C. Muller and M. Radoui, XXIIème Journées de Statistiques Tours, Réseaux neuromimétiques et analyses des données, Electricité de France (E.D.F Clamart) (1990).
3. R. Sobral, S. Canu, 71ème Congrès des AssociationsGénérales des hygiénistes et techniciens municipaux Annecy 15-19 avril 1991, textes de conférences pp 399-409, Application des réseaux de neurones artificiels à la prévision : la consommation d'eau (1991).
4. S. Thiria, LRI Université de Paris XI and Cedric CNAM, Réseaux de neurones : liens avec les techniques statistiques (1991-1992).
5. R. Tomassone, E. Lesquoy and C. Mullier, Inria Masson, page 24, La régression nouveaux regards sur une ancienne méthode statistique (1983).
6. F. Badran, S. Thiria, F. Fogelman Soulie, Kadratoff et Diday Editeurs, Comparaison analyse des données et réseaux multicouches : Introduction symbolique et numérique à partir de données (1991).
7. H. Bourlard, Y. Kamp, Manuscript M217 Philips Research Lab, Auto-association by multilayer perceptrons and singular value decomposition, (1987).
8. H. Bourlard, Y. Kamp, in IEEE 1st ICNN San Diego, Multilayer perceptrons and automatic speech recognition, (1987).
9. H. Bourlard, C.J Wellekens, Philips Research Laboratory Belgium, Links between markov models and multilayers perceptrons (1989).
10. G. Dreyfus, L. Personnaz, Journées Internationales des Sciences Informatiques Tunis, Intriduction au réseaux de neurones formels (1990).
11. P. Gallinari, F. Fogelman Soulie, S. Thiria, 2nd annual international conference on neural network classifiers San Diego, Multilayer perceptrons and data analysis "Neural Networks for computing" (1988).
12. M.B Beck, IIE Proceedings Vol. 133, Pt. D, n° 5, Identification, estimation and control of biological waste-water treatment processes" (1986).
13. J.C Bezdek, Vol 1 n° 1 pp. 1-6, "Editorial - fuzzy models - What are they and why ?", IEEE Transactions on Fuzzy systems (1993).
14. G. Block, Spécialité Automatique, Modélisation des procédés : des méthodes, un outil logiciel", Thèse - Doctorat de l'Université de Nancy I (1988).
15. D. Filev, Vol. 5 pp 281-290, Fuzzy modeling of complex systems, International Journal of Approximate reasoning (1991).
16. B. Kosko, IEEE Int. Conf. on Fuzzy Systems, San Diego pp. 1153-1162, Fuzzy systems are universal approximators.
17. L. Ljung, Prentice Hall, System identification. Theory for the user (1987).
18. E.H. Mandani, Proc. IEE, vol. 121 pp. 1584-1588, Application of fuzzy algoritms for control of simple dynamic plant (1974).
19. M. Sugeno, G.T Kang, Fuzzy sets and systems, 28 pp. 15-33, Structure identification of a fuzzy model" (1988).
20. T. Tagaki, M. Sugeno, IEEE Transactions on systems, Mann and Cybernetics, 15, pp.116-132, Fuzzy identification of systems and its application to modelling and control (1985).
21. E. Tan, G. Mourot, D. Maquin, J. Ragot, CNRS URA 821, Identification of fuzzy models, Centre de Recherche en Automatique de Nandy (1994).
22. L.X WANG, IEEE Int. Conf. on Fuzzy Systems, San Diego, CA pp.1163-1170, Fuzzy systems are universal approximators (1992).
23. L.A Zadeh, Information and Control, Vol. 8 pp 338-353, Fuzzy sets.

24. M. Jauzein, Gip Stelor RH 96-08 Ministère de l'Environnement, Rapport bibliographique (1996).
25. M. Jauzein, Gip Stelor Ministère de l'Environnement, Rapport final en cours d'édition (1997).
26. P. Rocklin, Thèse pour présentation de doctorat INPL, Conception de sondes de détection et dévleoppement de techniques de mesure in situ de la contamination de matrices solides par des composés organiques volatils (1996).

PROCESS FLOW RATES RECONCILIATION AND DATA ANALYSIS

D. MAQUIN and J. RAGOT, Centre de Recherche en Automatique de Nancy, Ecole Nationale Supérieure de Géologie, 2, Avenue de la Forêt de Haye, 54 500 Vandoeuvre les Nancy, France

1. Abstract

For plant control improvement, coherency of information supplied by instrument lines and sensors must first be ensured. Because of the presence of random and possibly gross errors, the model equations of the process are not generally satisfied. The problem of how to reconcile the measurements so that they satisfy the model constraints and how to use the reconciled values to detect faulty measurements (often called gross errors) are considered in this article. The authors strive to summarise various aspects of data reconciliation with the double aim of presenting the state of the art and bringing out the major difficulties encountered in the field. For that, the different steps of the methodology are presented in the following order : data reconciliation techniques, gross errors detection and gross errors localisation.

2. Introduction

The safety of processes can be greatly enhanced through the detection and isolation of the changes indicative of modifications in the process performances. If the models describing the process are accurate, the problem of fault detection may be solved by observer-type filters. These filters generate the so-called residuals computed from the inputs and the outputs of the process. This residual generation is the first stage in the problem of fault detection and identification (FDI). For them to be useful in FDI, the residuals must be insensitive to modelling errors and highly sensitive to the failures under consideration. In that regard, the residuals are designed so that the effects of possible failures are enhanced which in turn increases their detectability. The residuals must also respond quickly. The second stage of FDI is concerned with residuals analysis and decision making; the residuals are examined for detecting the presence of failures. The use of simple decision rules such as threshold tests or more sophisticated approaches using pattern recognition, sequential probability ratio test or sensitivity analysis is very helpful at this stage. Various FDI methods have been reported in the literature, notably in the excellent survey papers of Willsky (1976), Iserman (1984), Frank (1990), Gertler (1991). Among the classical books on the subject are those of Pau (1981), Basseville (1986a), Patton (1989b), Dubuisson (1990).

It should be noted that in the field of chemical and mineral processing industry, data reconciliation using material balance steady-state models has been extensively studied and applied. The particular point of view of data reconciliation and gross error detection is addressed in the book of Ragot (1990a). In the last two decades, the problem of detecting and identifying gross errors in measurements due to biases in the measuring instruments has been well studied. Several statistical tests for this purpose have been developed, such as for example, the constraint test (Mah and al., 1976), the measurement test (Mah and Tamhane, 1982), the iterative measurement test method, the screen combinatorial method (Serth, 1986), the dynamic measurement test (Rosenberg, 1987), the generalised likelihood ratio test (Narasimhan, 1987) or the maximum power test (Crowe, 1989). There are different ways to identify a large error : with a theoretical analysis of all effects leading to this error, with hardware redundancy by measuring a given process variable with different sensors, by checking the consistency of the raw data. This third alternative is selected here ; it is based on analytical redundancy by using the model equations of the process.

The general procedure of error detection is classically divided into two main parts : firstly the generation of so-called residuals, which are functions of measurements that are accentuated by the errors, secondly the detection, the isolation and the estimation of the error. These two points will be presented in this paper and an example is provided in order to illustrate the gross error detection technique.

3. Static redundancy equations

Historically speaking, likely due to measurement availability, static redundancy equations have first been utilised in the mineral processing and chemical industries as well as for electrical distribution networks. The first studies (Ripps 1962, Vaclavek 1969) were concerned with data reconciliation using the now classical technique of equilibration of production's balances. In the following stages this data reconciliation principle has been generalised to processes which are described by algebraic equations either linear (Crowe 1983) or non linear (Sood 1979, Crowe 1986). At the same time, data reconciliation went into use for more general applications than establishing statistically coherent balances.

3. 1. Principles of redundancy generation

The linear relationship between the measurements Y and the actual values X of a process variables vector is given in a simple matrix form as:

$$Y = CX + e \quad (1)$$

where Y is the (c.1) data vector whose entries are obtained from either sensors or analytic relationships, C the (c.m) measurement matrix, X is the (m.1) actual values of the process variable vector and e the (c.1) noise vector associated to the data. It is assumed that the noise is zero mean and characterised by a known variance matrix V (which is diagonal if the measurement errors are independent).

For obtaining the vector X, a minimum m, out of the c measurements, is needed. Therefore, redundancy in measurements generally (sufficient condition) appears when the inequality $c > m$ holds.

Data inconsistency can be easily pointed out by eliminating the unknown variables X from equation (1) when e is null. This yields $(c-m)$ linearly independent equations known as parity equations. Potter and Suman (1977) have established a general formulation of this problem when the covariance matrix of the measurement errors is equal to unity. When the covariance matrix is not equal to unity, let us define the $((c-m).c)$ projection matrix W so that it fulfils several conditions among them :

$$WV^{-1/2}C = 0 \quad (2)$$

Under these conditions, the redundancy equations can be written under the following form:

$$WV^{-1/2}Y = 0 \quad (3)$$

When the hypotheses about measurement noises are not valid, different situations can appear. In the case of errors whose mean values are not equal to zero, because of a systematic error of a sensor, the problem can easily be solved. Indeed, this systematic error can be statistically estimated which allows a correction of the measurements. For this estimation to be performed, the faulty measurement has first to be localised and identified. Many authors have studied this difficult localisation problem in the case of static redundancy equations. The main localisation methods will be presented in the third section of this chapter.

3. 2. The constrained case. A systematic decomposition

Let us now consider the system described by a linear constraint and a measurement equation:

$$AX = 0 \quad (4a)$$
$$Y = CX + e \quad (4b)$$

X is a v-dimensional state vector, Y is a m-dimensional vector of measured outputs, A and C are known matrices of appropriate dimensions.

Using simple transformations this equation can be reduced to those used for the unconstrained case. Linear systems, in which redundancy is present, can be written under various forms depending on the structure of the constraint and measure equations. However, through a few simple transformations they can be reduced to a unique representation defined by equations (4) or under an equivalent form:

$$\begin{pmatrix} I \\ 0 \end{pmatrix} Y = \begin{pmatrix} C \\ A \end{pmatrix} X + \begin{pmatrix} I \\ 0 \end{pmatrix} e \quad (5)$$

Then after a slight rearrangement, we can consider the same case that one given by (1) and therefore the generation of redundancy equations may be performed by the same way.

3. 3. Example

The procedure can be illustrated using the following simplistic example. In this case the system is described by the measurement equation (1) without measurement errors. It is a priori redundant as there are 4 observations of only 2 variables.

$$Y = \begin{pmatrix} 1 & 2 \\ 1 & 0 \\ 1 & 1 \\ 2 & 0 \end{pmatrix} X$$

It is easy to solve (2) with V reduced to an identity matrix. Here, with regard to the simplicity of the measurement equation, a direct elimination procedure can also be applied. Then, the redundancy equations are expressed :

$$\begin{pmatrix} 1 & 1 & -2 & 0 \\ 0 & 4 & 0 & -2 \end{pmatrix} Y = \begin{pmatrix} 0 \\ 0 \end{pmatrix} \quad (6)$$

When the measurements are affected by faults, the redundancy equations are not fulfilled and this may be enhanced by defining the residuals :

$$\begin{cases} r_1 = y_1 + y_2 - 2y_3 \\ r_2 = 4y_2 - 2y_4 \end{cases} \quad (7a)$$

If the main goal is to detect measurement faults, these residuals are compared to threshold values. Examination of the residual structure shows that faults on y_1 and y_3 are detectable but non isolable while they belong to the same residual equation. More generally, detectability and isolability mainly depend on the structure of the equations and on the occurrences of the variables in each of them. This may be analysed more deeper by considering all the redundancy equations ; for example, from (6), it is possible to eliminate the common variable y_2 which gives a third residual :

$$r_3 = -4y_1 + 8y_3 - 2y_4 \quad (7b)$$

Thus, considering (7) show that each residual depends on a certain number of variables and therefore specific signature of gross errors may be generated according to the occurrence of the variables.

4. Dynamic redundancy equations

The different concepts that have so far been introduced for studying the observability of linear systems can also be perfectly applicable to the case of dynamic systems. In particular the previously defined classification of variables according to their deducibility and redundancy can be generalised. The only difficulty lies in the increase in dimension of the vector of variables which is a function of the dimension of the observation time-window. Because of this similarity between the static and dynamic case, fault detection (sensors and actuators) can be achieved using the principles defined in the case of static redundancy equations. In a practical sense, it is possible to generate redundancy equations from state equations either in time or frequency domain. These two approaches are totally equivalent (Lou, 1982) if some conditions of duration of the observation time-window are satisfied. Whatever the approach, the basic principle is the same: the unknown variables are eliminated so that only known or measured variables appear in the equations.

4. 1. Presentation

Consider the following deterministic model (8) where x is the n-dimensional state vector, A is a n.n matrix, B a n.p matrix, C a mp.n matrix. The vectors u and y correspond to the known inputs and outputs of the process. In all the following treatments, without loss of generality, the measurement y depends only on the state x and do not include the input u.

$$x(k+1) = Ax(k) + Bu(k) \quad (8a)$$
$$y(k) = Cx(k) \quad (8b)$$

Direct redundancy may exist among sensors whose outputs are algebraically dependent; this corresponds to the situation where the variable measured by one sensor can be determined instantaneously by other sensor measures. This direct redundancy is very useful for sensor failure detection but is not applicable for the detection of actuator failures. In this situation, the temporal redundancy which links sensor outputs and sensor inputs must be established. When integrated on $[k,k+r]$ window, the system (8) is expressed as:

$$Y(k,r) - G(r)U(k,r) = H(r)x(k) \quad (9)$$

where Y is the $m(r+1)$ vector of the outputs $y(k)$ to $y(k+r)$, U is the pr vector of the inputs $u(k)$ to $u(k+r)$, G a $m(r+1).(pr)$ matrix and H a $m(r+1).n$ matrix; $H(r)$ is called the r-order observability matrix of the process. Thus, it is clear that the structure of (9) is the same that those of equation (1).

4. 2. Example

Let us consider an example taken from Massoumnia (1988) with state matrices:

$$A = \begin{pmatrix} 0 & 0 & 0 \\ 0 & 2 & 0 \\ 0 & 0 & 1 \end{pmatrix} \quad B = \begin{pmatrix} 1 & 0 \\ 0 & 1 \\ 1 & 1 \end{pmatrix} \quad C = \begin{pmatrix} 1 & 1 & 0 \\ 0 & 0 & 1 \end{pmatrix}$$

Direct calculation of the output system on the window $[k,k+2]$ according to (8) and elimination of the state yields the two redundancy equations:

$$2y_1(k+1) - y_1(k+2) - 2u_1(k) + u_1(k+1) + u_2(k+1) = 0$$
$$-y_2(k+1) + y_2(k) + u_1(k) + u_2(k) = 0$$

Assuming that the actuators are fully reliable, the first equation is sensitive to the failure of the first sensor, and the second one is sensitive to the failure of the second sensor. Hence these two equations can be used to identify any sensor failure. By aggregation of the two equations, considering u_1 or u_2 as common variables, we obtain:

$$y_1(k+2) - 2y_1(k+1) - y_2(k+2) + 2y_2(k+1) + u_1(k) = 0 \quad (10a)$$
$$-y_1(k+2) + 2y_1(k+1) + y_2(k+2) - 3y_2(k+1) + 2y_2(k) + 2u_2(k) + u_2(k+1) = 0 \quad (10b)$$

Assuming that the sensors are fully reliable, the first equation is sensitive to the failure of the first actuator and the second equation is sensitive to the failure of the second actuator. Hence these two equations can be used to identify any actuator failure.

5. Residuals analysis

The first two sections were devoted to the generation of redundancy equations for linear static and dynamic systems. As already mentioned, the second stage of FDI concerns the so-called residuals evaluation i.e. the forming of diagnostic decision on the basis of the residuals. To limit the length of the present paper, all the aspects of this stage will not be covered in this section. We will especially focus on the methods issued from static systems analysis and which can be extended to dynamic systems (Maquin, 1991b). As indicated by Gertler (1990), the decision making stage usually implies statistical testing. There is a close relationship between statistical testing and residual generation. Residuals are variables that are zero under ideal circumstances; they become nonzero as a result of failures, noise and modelling errors. To account for the presence of noise, statistical testing is applied to the residuals. Then, a logical pattern is generated showing which residuals can be considered normal and which ones indicate faults. Such a pattern is called the signature of the failure. The final step of the procedure is the analysis of the logical patterns obtained from the residuals. The aim is to isolate the failures. Such analysis may be performed by comparison to a set of signature each of them representing a given failure.

5. 1. Presentation of the process model

In this section, residuals generated either from static or dynamic systems are in a unified framework. A linear system can be described, in the fault-free case, by the following model and measurement relations:

$$MX = 0 \quad (11a)$$
$$Z = HX + e \quad (11b)$$

where X is the v-dimensional vector of process variables, Z the m-dimensional vector of measurements, M the n.v matrix of model equations (without loss of generality, it is supposed of full row rank), H the m.v measurement selection matrix and e is a vector of random errors characterised by its variance matrix.

For dynamic processes, the model, which relates the state vector $x(k)$ to the actual input vector $u(k)$ and the output vector $y(k)$, described in state space discrete form, may be written as :

$$x(k+1) = Ax(k) + Bu(k) \quad (12a)$$
$$y(k) = Cx(k) \quad (12b)$$

where $u(k)$ and $y(k)$ denote the actual values of the input and output of the system. Defining, on a time-window of length N, the mixed vector of inputs and states and the corresponding constraints matrix:

$$X = \begin{pmatrix} x(0) & u(0) & x(1) & u(1) & \cdots & u(N) & x(N+1) \end{pmatrix}^T \quad (13)$$

$$M = \begin{pmatrix} A & B & -I & . & . & . & . & . \\ . & . & A & B & -I & . & . & . \\ . & . & . & . & . & . & . & . \\ . & . & . & . & . & A & B & -I \end{pmatrix} \quad (14)$$

the constraint equation and the measurement equations (11) may be condensed into the form:

$$MX = 0 \quad (15a)$$
$$Z = HX + e \quad (15b)$$

As the inputs and only a part of the state (15b) are measured, the selection matrix H is defined by:

$$H = \begin{pmatrix} C & 0 & & & \\ 0 & I & 0 & & \\ 0 & 0 & C & & \\ . & & & . & \\ . & & & & C \end{pmatrix} \quad (16)$$

As equations (15a) and (15b) are strictly identical to (11a) and (11b), static and dynamic systems can be analysed in a unified framework.

5. 2. State estimation

The methods for fault detection and isolation are often divided into two groups: those which apply a priori, without carrying out the full data reconciliation (estimation of process variables), by directly testing the residuals issued from redundancy equations and those which apply a posteriori on the residuals generated by calculating the differences between the raw measurements and their estimations. In practice, both methods are used together in order to improve the robustness of the procedure of fault detection and isolation.

The estimation or data reconciliation problem of system (11) involves finding a set of adjustments such that the adjusted values verify the model equation (11a). With the classical assumption that the measurement errors e are normally distributed with zero mean and known variance matrix V, this optimisation problem can be stated as maximising the probability density function (Ragot, 1990a):

$$P(Z)=\frac{1}{(2\pi)^{m/2}|V|^{1/2}}\exp\left(-\frac{1}{2}(Z-HX)^T V^{-1}(Z-HX)\right) \quad (17)$$

subject to $MX=0$

Using the Lagrange multipliers technique leads to the classical unbiased estimator:

$$\hat{X}=\left(G^{-1}-G^{-1}M^T(MG^{-1}M^T)^{-1}MG^{-1}\right)H^T V^{-1}Z \quad (18a)$$

$$G=H^T V^{-1}H+M^T M \quad (18b)$$

Due to space limitation, the following analyses focus on the case corresponding of the measurement of all process variables, i.e. H is equal to an identity matrix. This situation is reducing but it allows the main ideas of the analyses to be presented. For a complete description of the proposed methods, the reader is referred to (Ragot et al., 1990). In this simplified case, the previous estimator may be written under the form:

$$\hat{X}=\left(I-VM^T(MVM^T)^{-1}M\right)Z \quad (18c)$$

The vector E of adjustments (or residuals) and the residual criterion ϕ_R are then obtained by direct substitution:

$$E=Z-\hat{X}=VM^T(MVM^T)^{-1}MZ \quad (19a)$$

$$\phi_R=E^T V^{-1}E \quad (19b)$$

Both the vector of adjustments E (19) and the direct imbalances vector of redundancy equations R defined by :

$$R=MZ \quad (20)$$

can be considered and processed as residuals. However, it should be noticed, that these residuals cannot be analysed in the same way as each entry of R is associated with an equation and each entry of E with a specific variable.

5. 3. Residual criterion analysis

A first approach to testing the residuals is to introduce a single scalar statistic like, for example, the residual criterion (19). As was first pointed out by Reilly (1963), the residual criterion ϕ_R has a chi-square distribution with a number of degrees of freedom equal to the rank of M. Furthermore, it is also useful to note that the calculation of ϕ_R does not require the estimation stage. It is easy to show that the residual criterion can be expressed as a function of R :

$$\phi_R = R^T \left(MVM^T\right)^{-1} R \quad (21)$$

Thus the residuals can be globally tested against tabulated values of chi-square. In the fault-free case, the function ϕ_R is below the threshold for the chi-square with the appropriate confidence level and number of degrees of freedom. Unfortunately, if the chi-square test is satisfied, it does not prove that there are no faults in the measurements set since a fault may exist among a large set of measurements. It is then preferable to use further specific tests to diagnose the measurements.

A difficulty with this global test is that, while it indicates well the presence of fault it is not able to identify the source of these errors. The use of a sequential procedure allows the location of the fault (see section 6).

5. 4. Imbalances or adjustments vectors analysis

Another approach is the direct parallel testing of the residuals. With the assumption of a Gaussian distribution of the measurement errors, the vector R also follows a normal distribution with zero mean and covariance V_R:

$$V_R = MVM^T \quad (22)$$

In order to compare the elements of the R vector, let us define a standardised imbalance vector R_N :

$$R_N = diag(V_R)^{-1/2} R \quad (23)$$

Each entry $R_N(i)$ follows a normal distribution with zero mean and unity variance. A simple statistical two tailed test can therefore be used: we may conclude that equation i is a “bad” equation if:

$$|R_N(i)| > t \quad (24)$$

Classically, one may choose the critical constant t to control the familywise Type I error rate at some pre-assigned level a. Even if we assume the presence of only one gross error, the relationship between the “bad” equation(s) and the suspect measurement is not

straightforward. It depends on the structure of the equations and the location of the faults. In some cases, we are not able to suspect one measurement only.
For solving this case (Mah, 1976) proposed to apply the preceding test to each equation and also to the aggregates of two or more equations (also known as pseudo-equations). The main assumption underlying this method is that faults do not cancel each other.

This latter approach can also be applied to the adjustments vector E. The variance matrix of this vector is expressed as:

$$V_E = VM^T(MVM^T)^{-1}MV \quad (25)$$

As for the imbalance residuals vector, we define the standardised adjustments vector:

$$E_N = diag(V_E)^{-1/2}E \quad (26)$$

Each $E_N(i)$ is compared with a critical test value. If at least one entry of E_N is out of the confidence interval then, there is a "bad" measurement. The defective measurement can always be shown to correspond to the greatest standardised adjustment residual (Fayolle, 1987).

For the linear case, instead of (26), Tamhane (1985) has shown that for a non diagonal covariance matrix V, a vector of test statistics with the maximal power for detecting a single fault is obtained by premultiplying E by V^{-1}. Then, the transformed residual, $e = V^{-1}E$, is normally distributed with zero mean and a variance matrix $V_e = V^{-1}V_EV^{-1}$. The power of the test (the probability of correctly detecting and identifying gross errors when they are present in the process data) has been established and discussed by Iordache (1985) under different conditions (various networks, errors location, variance values ...) using the Monte Carlo simulation.

6. Sequential technique for gross error detection

The use of residual criterion analysis coupled with a sequential procedure of measurement elimination allows one to locate the faulty measurements. For the set of all process measurements one first calculates the residual criterion ϕ_R ; if an error is indicated by the global test, all measurements are considered as suspect candidates. Then, the measurements are "deleted" sequentially from the process (in groups of size 1, 2, ...). After each deletion the global test is again applied. In this approach we wish to assess the effect of deleting a particular set of measurement on the objective function and on the estimations. This scheme, firstly proposed Ripps (1962), was also used by Nogita (1972) under a slightly modified form. Moreover, it is possible to have the same approach as the one developed in the case of multiple-observer for state reconstruction (Frank, 1989) by comparing together the different estimations obtained after each deletion. It is also possible to consider suspect measurements by assigning them an infinite variance. The corresponding variation of the criterion f is then used to detect the possible gross errors.

Let us consider the following example :

$$\begin{cases} x1 + x2 - x3 = 0 & \text{(equation 1)} \\ x3 + x4 - x5 = 0 & \text{(equation 2)} \\ x5 + x6 - x2 - x7 = 0 & \text{(equation 3)} \\ x7 - x8 - x9 = 0 & \text{(equation 4)} \end{cases}$$

The following table gives the values of measurements and their respective standard deviations.

Table 1. Measurements and their accuracies

Variable	**1**	**2**	**3**	**4**	**5**	**6**	**7**	**8**	**9**
Measurement	111.3	18.2	191.4	23.8	148.7	13.6	181.2	106.4	39.7
Standard deviation	2.8	0.5	3.2	0.6	3.7	0.3	3.5	2.7	1

All the variables are assumed to be measured. Practically, if it is not the case, the redundant part of the system must be extracted.

The table 2 shows the considered equations, the corresponding residual values, the normalised residuals (residuals divided by their standard deviations), the nature of the residual (normal or abnormal), the threshold being fixed to 2 and the variables involved in the calculus of the residual.
The last eight rows of the table correspond to residuals which are issued from the aggregation (sum) of some primary equations.The analysis of this table is based on the following remark. If the residual of an equation is not abnormal, the measurements used in the calculus of this residual are not contaminated by gross errors. This assertion lies on the hypothesis of simultaneous occurrence of only one gross error (elsewhere some compension may occur).

So, for the considered example, the measurements which intervene in the calculus of equations 1+2, 3+4, 1+2+3+4 are fault free. Then, it is easy to conclude that the remaining measurements x_3 et x_7 are probably faulty.

Table 2. Analysis of residuals

Equation	**Residual**	**Normalised residual**	**Abnormal residual**	**Variables**
1	-61.9	-14.5	yes	1,2,3
2	66.5	13.4	yes	3,4,5
3	-37.1	-7.2	yes	2,5,6,7

Equation	Residual	Normalised residual	Abnormal residual	Variables
4	35.1	7.8	yes	7,8,9
1+2	4.6	0.9	no	1,2,4,5
1+3	-99.0	-14.9	yes	1,3,5,6,7
2+3	29.4	6.1	yes	2,3,4,6,7
3+4	-2.0	-0.4	no	2,5,6,8,9
1+2+3	-32.5	-7.2	yes	1,4,6,7
1+3+4	-63.9	-10.1	yes	1,3,5,6,8,9
2+3+4	64.5	14.8	yes	2,3,4,6,8,9
1+2+3+4	2.6	0.6	no	1,2,6,8,9

This error detection-isolation procedure may be implemented as follows.The boolean value "true" coded by the value "1" is affected to the variables which intervene in an abnormal residual and the boolean value "false" (0) is affected to the variables which intervene in a normal residual. When the variable do not intervene in the considered equation, the symbol "." is used. The last row of the table 3 is the logical product of the boolean values in a column (in this calculus, the symbol "." is not taken into account). The elements equal to "1" in this last row indicate the suspect measurements. Notice that the aggregation of equations may be achieved sequentially as well as the logical product, so the isolation may be obtained before doing all the possible combinations.

This technique may be extended to dynamical equations or non-linear equations. However, elimination of variables (aggregation of equations) in the non-linear case may be more difficult and may required the usage of specific techniques.

Table 3. Logic analysis

Equation	Abnormal residual	1	2	3	4	5	6	7	8	9
1	yes	1	1	1	.	.	.	.	.	.
2	yes	.	.	1	1	1	.	.	.	.
3	yes	.	1	.	.	1	1	1	.	.
4	yes	.	.	.	.	.	.	1	1	1
1+2	no	0	0	.	0	0	.	.	.	.
1+3	yes	1	.	1	.	1	1	1	.	.
2+3	yes	.	1	1	1	.	1	1	.	.
3+4	no	.	0	.	.	0	0	.	0	0
1+2+3	yes	1	.	.	1	.	1	1	.	.
1+3+4	yes	1	.	1	.	1	1	.	1	1
2+3+4	yes	.	1	1	1	.	1	.	1	1
1+2+3+4	no	0	0	.	.	.	0	.	0	0
Product		0	0	1	0	0	0	1	0	0

7. Architecture of a measurement validation system

The different stages of the measurement validation of a process are depicted in figure 1. The different sources of information allowing the analysis are shown : the process model, the location of sensors, the values and the accuracies of measurements and the reliabilities of sensors. These information may be used in different levels : structural data (process model and location of sensors) are used in observability and redundancy analysis of the variables. The reliabilities of sensors may be useful for introducing failure rate of sensors and analysing the whole reliability of the instrumentation scheme. The values and the precisions of the measurements are used by state estimation and residual generation. The analysis of measurements for fault detection and gross error localisation may be done a priori (analysis of the residuals generated from the process model) and/or a posteriori (analysis of the residuals generated by comparison of the measurements and their estimations).

The bold lines indicate an eventual modification stage of the measurement validation system. Indded, when the fault detection-isolation is not possible, the number,the location, the accuracies or the reliabilities of sensors must be modified. This problem may be expressed as a constrained optimisation problem ; the constraints taking into account the redundancy of variables and the fault detectability and isolability (Maquin, 1995, 1997).

Figure 1. Architecture of a measurement validation system

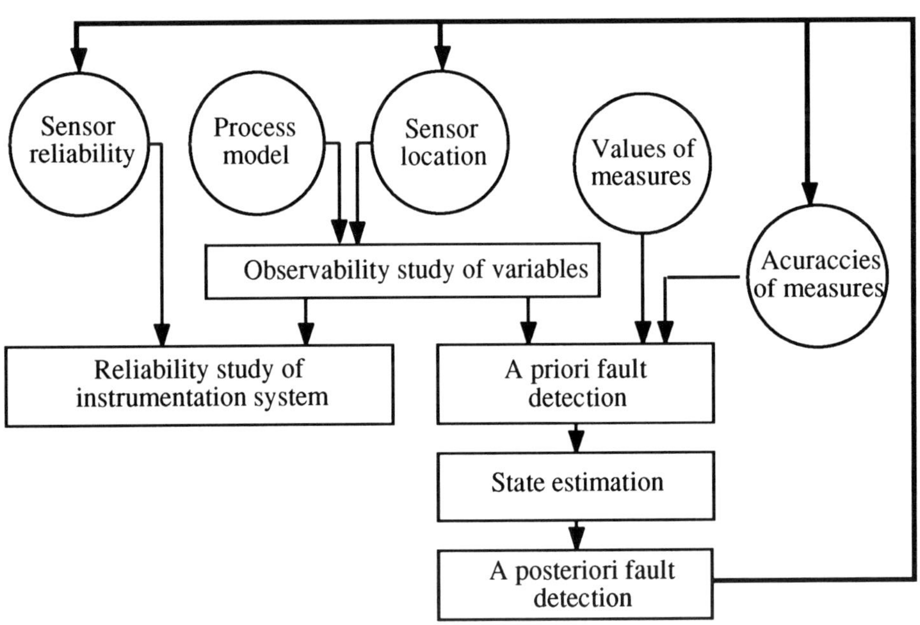

8. Conclusion

One of the main difficulties in decision making using plant data is that true values for stream flow are not known. Because of perturbations in environmental conditions and malfunctions in equipment and instrumentation, its becomes necessary to adjust or rectify the process data. Analytic redundancy among process variables have been used to filter out measurement noise in data reconciliation. Throughout this communication, the authors have presented a short overview of the important problem of data validation in the linear case. We have shown that redundancy can filter not only the measurement noise but also gross errors due to sensor misfunctionning. However, we have to point out, that the implementation of such techniques need the knowledege of the process model. It is obvious, that the robustness of the data reconciliation techniques have to be improved in regard of uncertainties of the model. Probably, techniques taking into account parameters uncertainties, such as fuzzy modelization or interval arithmetic, may be used to reduced the effects of a bad modelization.

REFERENCES

1. Basseville M., Nikiforov I.V. Detection of abrupt changes - Theory and applications. Series in Information and System Sciences, Prentice-Hall, 1993.
2. Broussaud A., Clin F., Croise G., Fourniguet G. Méthodologie d'échantillonnage et bilans matière cohérents dans un pilotage de minerai sulfuré. Revue de l'Industrie Minérale, p. 165-170, 1988.
3. Brunet M., Jaume D., Labarrère M., Rault A., Vergé M. Détection et diagnostic de pannes. Traité des nouvelles technologies, série diagnostic et maintenance, Hermès, 1990.
4. Cassar J.P., Staroswiecki M. Advanced design of the decision procedure in failure detection and isolation systems. IFAC Symposium on Fault Detection, Supervision and Safety for Technical Processes, SAFEPROCESS'94, Helsinki, Finland, June 13-16, 1994
5. Chow E.Y., Willsky A.S. Analytical redundancy and the design of robust failure detection systems. IEEE Trans. on Automatic Control, vol. AC-29, p. 603-614, 1984.
6. Clark R.N. The dedicated observer approach to instrument failure detection. 18th IEEE Conference on Decision and Control, p. 237-241, 1979.
7. Crowe C.M. Observability and redundancy of process data for steady-state reconciliation. Chemical Engineering Science, vol. 44, n° 12, p. 2909-2917, 1989.
8. Dubuisson B. Diagnostic et reconnaissance de formes. Traité des nouvelles technologies, série diagnostic et maintenance, Hermès, 1990.
9. Fayolle J. Validation de données et diagnostic des procédés industriels. Thèse de Doctorat de l'Université de Nancy I, septembre 1987.
10. Frank P.M. Fault diagnosis in dynamical systems using analytical and knowledge based redundancy - a survey and some new results. Automatica, vol. 26, n°3, p. 459-474, 1990.
11. Gertler J. Analytical redundancy in fault detection and isolation. IFAC Symposium on Fault Detection, Supervision and Safety for Technical Processes, SAFEPROCESS'91, 9-22, Baden-Baden, Germany, 1991.

12. Kratz F., Héraud N., Ragot J. Diagnostic de systèmes automatiques par équations analytiques de redondance. Revue d'Automatique et de Productique Appliquées, vol. 5, n° 1, pp 25-48, 1992.
13. Kratz-Bousghiri S., Nuninger W., Kratz F. Fault detection in stochastic dynamic systems by data reconciliation. Journal of Engineering Simulation, special issue on Engineering Diagnostics, G.E. Pukhov ed., vol. 17, n° 5, 1995.
14. Kratz F., Bousghiri S., Nuninger W. A finite memory observer structure of continuous descriptor systems. American Control Conference, Seattle, USA, June 21-23, 1995.
15. Lou X.C., Willsky A.S., Verghese G.C. Optimally robust redundancy relations for failures detections in uncertain systems. Automatica, vol. 22, n°3, p. 333-344, 1986.
16. Luong M., Maquin D., Huynh C.T., Ragot J. Observability, redundancy, reliability and integrated design of measurement system. 2nd IFAC Symposium on Intelligent Components and Instruments for Control Applications, SICICA'94, Budapest, Hungary, June 8-10, 1994.
17. Luong M., Paris J.M., Maquin D., Ragot J. Observability, reliability and sensor positioning. 1995 AIChE Spring National Meeting, Houston, USA, March 19-23, 1995.
18. Mah R.S.H., Stanley G.M., Downing D.M. Reconciliation and rectification of process flow and inventory data. I.E.C. Process Design Development, vol. 15, n°1, p. 175-183, 1976.
19. Maquin D., Fayolle J., Darouach M., Ragot J. Le positionnement de capteurs dans les systèmes industriels de grande dimension. Proceedings of IMACS congress, Villeneuve d'Ascq, 1986. Published in Applied Modelling and Simulation of Technological Systems, Elsevier, 1987.
20. Maquin D., Ragot J. LVD : a software for mass balance and data validation. Dechema-Monographs vol. 116, VCH Verlagsgesellschaft, 1989.
21. Maquin D., Bloch G., Ragot J. Data reconciliation of measurements. Revue Européenne Diagnostic et Sûreté de Fonctionnement, vol. 1, n°2, p. 145-181, 1991.
22. Maquin D., Ragot J. Comparison of gross errors detection methods in process data. 30th IEEE Conference on Decision and Control, p. 2253-2261, Brighton, December 1991.
23. Maquin D., Gaddouna B., Ragot J. Parity space equations for singular systems. IEEE International Conference on Systems Man and Cybernetics, Le Touquet, 1993.
24. Maquin D., Luong M., Ragot J. About the design of measurement systems and fault accommodation. Journal of Engineering Simulation, special issue on Engineering Diagnostics, G.E. Pukhov ed., vol. 17, n° 5, 1995.
25. Maquin D., Luong M., Ragot J. Some ideas about the design of measurement systems. European Control Conference ECC'95, Roma, Italia, September 5-8, 1995.
26. Maquin D., Luong M., Ragot J. Fault detection and isolation and sensor network design. European Journal of Automation, vol. 31, n° 2, p. 393-406, 1997.
27. Narashiman S., Mah R.S.H. Generalized likelihood ratio method for gross error identification. AIChE Journal, vol. 33, n° 9, p. 1514-1521, 1987.
28. Nogita S. Statistical test and adjustment of process data. Industrial Engineering Chemistry Process Design Development, 11 (2), p. 197-200, 1972.
29. Potter J.E., Suman M.C. Thresholdless redundancy management with arrays of skewed instruments. Electronic flight control systems, Agardograph 224, p. 2115-2121, 1977.
30. Ragot J., Darouach M., Maquin D., Bloch G. Validation de données et diagnostic. Traité des nouvelles technologies, série diagnostic et maintenance, Hermès, 1990.

31. Ragot J., Maquin D., Bloch G. Sensor positionning for processes described by bilinear equations. Revue Européenne Diagnostic et Sûreté de Fonctionnement, vol. 2, n° 2, p.115-132, 1992.
32. Ragot J., Maquin D., Kratz F. Observability and redundancy decomposition - application to diagnosis. To be published in Advances in fault diagnosis for dynamical systems, Prentice-Hall, 1997.
33. Ragot J, Maquin D., Ribbens W.B. Two approaches for the isolation of plant failures in dynamic systems. International Conference on Fault Diagnosis, Tooldiag, p. 1-9, Toulouse, 1993.
34. Ragot J., Maquin D. Failure detection identification and accomodation based on a new accomodation filter. International Journal of Systems Science, vol. 24, n° 6, p. 1215-1220, 1993.
35. Ragot J., Maquin D. An algorithm for obtaining the redundancy equations of LTI systems. Automatica, vol. 30, n° 9, 1994.
36. Ripps D.L. Adjustment of experimental data. Chemical Engineering Progress, vol. 61, n° 55, p. 8-13, 1962.
37. Schreier G., Gaddouna B., Ragot J. Matrices computation of an unknown-input observer. International Journal of Systems Science, vol. 26, n° 4, p. 991-998, 1995
38. Sood M.K., Reklaitis G.W., Woods J.M. Solution of material balances for flowsheets modeled with elementary modules : the unconstrained case. AIChE Journal, vol. 25, n° 2, p. 209-219, 1979.
39. Staroswiecki M., Hamad M. Validation of measurements and detection of sensors failures in control systems. Signal processing III : theory and applications. I.T. Young editor, Elsevier, 1986.
40. Staroswiecki M., Cassar J.P., Cocquempot V. Generation of optimal structured residuals in the parity space. 12th IFAC World Congress, p. 505-510, Sydney, 1993.
41. Tamhane A.C., Mah R.S.H. Data reconciliation and gross error detection in chemical process network. Technometrics, vol. 27, n° 4, p. 409-422, 1985.
42. Turbatte H.C., Maquin D., Huynh C.T., Ragot J. Conception de l'architecture d'un système d'instrumentation. Application à un procédé pétrochimique. International Conference on Fault Diagnosis, Tooldiag, p. 406-415, Toulouse, 1993.
43. Vaclavek V. Studies on system engineering III. Optimal choice of the balance measurement in complicated chemical systems. Chemical Engineering Science, vol. 24, p. 947-955, 1969.

SENSOR NEEDS FOR WATER MONITORING

M. SCOTT, Secretary & Director, SWIG - Sensors for Water Interest Group, 27 West green, Barington, Cambridge, CB2 5RZ, United Kingdom.

1. Introduction

Measurement is easy but getting the correct result is difficult and convincing others even more so.

The market needs are :

- Higher confidence in the data; better reliability, reduced sensor fouling, better calibration audit trails, better sampling
- Lower skill requirements
- Lower cost of ownership
- Better integration of multi-variable data

To meet changing European Directives, competitive pressures and customer expectations all sectors of the water and waste treatment industries are trying to improve the monitoring of water quality. The number of water quality parameters to be measured is increasing rapidly and a greater precision is being demanded from existing sensors in the water and waste water industry (W&WT).

Technology is changing at an unprecedented speed, providing new opportunities and new challenges. However the W&WT industry has a small installed base of sensors and, in general, the users lack the resources to deal with any sophistication.

The data derived from the on-line analysers will increasingly be used by a wide range of users (clients) to make decisions in operations, business management and regulatory matters and this is a considerable challenge to both the manufacturers and the users.
The demand for rapid, accurate and simple methods of analysing key components in the W&WT industry, without the need to resort to costly, tedious, slow and error prone, sample to Laboratory method, will continue to grow. However potential users of the data rarely have any understanding of the dynamics of the sampling to Laboratory chain or the difficulties in obtaining a sample which accurately represents the process. This leads to unrealistic ideas as to the certainty which can reasonably be applied to the results of samples evaluated in a laboratory. It follows that unrealistic views are expressed about the performance required of measurements to be made in the field.

2. Market

It is possible, with little effort, to amass a bewildering amount of information on sensors for the UK water and waste treatment industry. It is in the nature of such accumulations that the reader becomes more confused as the information grows. This report, therefore, concentrates on the structure of the market sector and the issues that are seen to be the pacing factors for the introduction of new ideas and for the growth of established sensors. The statistics provided are used as a necessary basis for making macro judgements and as a starting point from which the reader can develop a specific market sector analysis for individual projects.

One of the problems for the application of measurement technology in the W&WT is that the European market is relatively small and probably declining. It is fairly easy to obtain estimates and forecasts for expenditure on Environmental protection equipment but it is less easy to obtain detailed information on the market sizes for the various monitoring market sectors. The OECD is a source for forecasts for environmental equipment some of which are reviewed in an article. (Ref. 1). On behalf of the UK department of Trade & Industry (Ref. 2), the author concluded, three years ago, that the UK market for water monitoring instruments was no more than 60M pa and anecdotal evidence suggests that it has decreased since then to probably 40M. The market for the water and waste treatment industry alone is certainly less. That 40m is made up of a number of sectors; for example test kits for water amount to no more than 6M and flow to no more than 20M and it is doubtful whether the market for Awet chemical analysers@ is more than three hundred units in the UK pa. It is traditional for market estimates to multiply the UK figures by seven or eight to obtain a European estimate (or divide the European figure by the same numbers!) But the result is still very small numbers. The problem is made more difficult since the number of measurands is large and increasing. It is easy to find wish lists of sixty or seventy measurands from the Water companies and the Regulators, with limits of detection which are close to or even smaller than current laboratory practice.

To provide a scale against which individual sensors market studies can be judged the chart provides estimates for the major sensor markets in the UK=s W&WT industry. Estimates of market size are completely dependent on definitions and the ability, or willingness, of suppliers to divulge useful numbers. The definition difficulty is shown in the overlap in flow; many level sensor sales are for open channel flow and not every supplier differentiates between types of flow measurement.
The "consumable" figure is even more difficult because it includes replacement probes as well as test kits and some companies treat replacement probes as base sales. Consumables may be overstated because of the difficulty of separating out, for example, test kits for private and municipal swimming pools. The non-quantified inclusion of Laboratory measurement is because many suppliers do not have a clear distinction between on-line sales and Laboratory sales. Some sample collecting and Laboratory procedures are often very close to the process and many users consider that they have on-line control. It is possible that the on-line sector is overstated by a few m. because of this overlap.

To provide another perspective of the potential market size it should be noted that the Water companies in the UK have 7,000 major sewage works consents and 21,000 combined sewer overflow consents and this when considered alongside the large equivalent industrial numbers this should represent a massive market. However very few of these measurement opportunities will be realised in the immediate future. Most of the publicly available market estimates, showing significant growth, are dependent on enforcement, by the regulators, of Legislation although a part of the drive will be because the water suppliers and waste treatment industries will need to operate more efficiently; the need to be a good neighbour may contribute a little to the market.

The available market statistics are distorted by events such as the need to protect against Crypotspiridium and Giardia; largely satisfied in the USA as well the UK with turbidity measurements which probably do not make an effective measurement for that application. The drive in some parts of Europe in the early '90s towards control of pollution by monitoring nutrient load also distorted the figures with a surge in the need for nitrate, phosphate and ammonia monitors. Once the need for a particular parameter is satisfied the maintenance levels of sales is very low. The most significant factor for anyone introducing or demanding a new measurement is the fact that there is no new money.

The market is driven by the need to meet and to demonstrate compliance with Legislation. Expenditure is therefore not seen as contributing to profits

The W&WT industry is extremely cost conscious and this combined with technological innovation, which is also driving down costs, will limit growth for existing products. Sales by numbers of units will grow significantly, but sales by value will be less buoyant unless the regulators force the adoption of new measurements. Substantial growth could come from new measurements but even with pressure from the regulators the user still has to find new money to fund increased expenditure.

3. Decision makers

The problem for the busy user is that legislative pressures and commercial demands impose a need for a considerably better understanding of how the business affects, and is affected, by the environment. The user therefore cannot avoid the challenges but often does not have the necessary time or knowledge to deal with new technology.

The decision making processes and the allocation of resources, for instrumentation, must usually come from a wide range of management groups and disciplines and the necessary co-ordination has often been missing. The result is that anecdotal stories abound about on-line analysers which have never worked. There is a current trend by the UK W&WT industry to remove all those analysers which make no contribution to the business and to properly resource those which are vital to the business. However this is a long process and it will be some time before the positive news dominates the black museum stories.

The result is that the supplier of sensor technology, whether a University supplying know how in some manner, or a manufacturer supplying a sensor, has difficulty locating the business and then has difficulty communicating to an often widely dispersed decision making mechanism.

It is suggested that a sound understanding of the business needs for a measurement and the circumstances of the use of the measurement device are more important than innovative technology.

Commercial success for the supplier is dependent on identifying the measurement need, selecting the appropriate format and incorporating the user's working conditions.

4. Cost of ownership

There is an increased awareness that the purchase cost of process sensors can represent a small part of the total cost of ownership and such factors as installation, calibration, validation and the cost of consumables of on-line instruments are receiving more attention. This is not a new consideration, but little progress has been made over several years of discussion and it is widely believed that the traditional purchasing structure, for new processes and plant, of the W&WT industry must be changed before "cost of ownership" becomes more important than "lowest price". The problem is that the sensor suppliers must inform a broad range of decision makers but they are rarely involved in the details of the final decision. More importantly, the supplier is rarely contracted to support the final user site. The Contractor receives an Order from the User based on design work by a Consultant and meets the terms of that Order at the lowest possible cost. There are many anecdotal stories of multiple suppliers for one particular measurand on a site because there has been no requirement on the Contractor to take existing measurements into consideration. There are also many anecdotal stories of analysers mounted in place but never properly commissioned because appropriate skills were not made available.

A major, and probably increasing, part of the W&WT market will be the supply of single or customised systems for specific applications. The supply chain may not be any simpler but the suppliers cost of selling as a percentage of Order value will be significantly greater and this will cause problems for both the supplier and the user.

The W&WT industry has yet to find a satisfactory structure for the allocation of resources, risk and reward for small monitoring applications.

5. Alternative technology

The major need for the W&WT industry is to show that organic and inorganic substances are only present, if at all, in quantities less than those determined by the prevailing regulations. This means that many measurements are at trace levels; certainly at low ppm, often at ppb and already down to ppt for some substances. Even Laboratory GCMS has problems quantifying herbicides at the EU required level of 0.1m/l.

Another problem is that different chemicals pose different risks for whole animals and chemical analysis must be an integral part of the risk assessment process for the business. Cocktails of chemicals may have different affects and individual chemicals may be themselves benign but cause other unwanted affects. The dilemma is that if you need to detect to very low levels then a GCMS laboratory analysis is needed; but are the assay techniques good enough to tell you when you need to take a sample and does the sample accurately represent the stream being monitored ?

It seems unlikely that the need for increased measurements will be met by increasing the number of single parameter on-line analysers; capital cost, cost of ownership and support resources would be a problem. True ownership includes purchase price, consumables, reagents, service and depreciation and might be as high as 5k per annum for each instrument and as already stated the number of chemicals, metals, organics and bacteria which need to be identified and measured at trace levels is increasing rapidly; a trend which can be expected to continue or even increase in the short and medium term future. There is room for improvement in the majority of the single measurand field measurements even including such common measurements as flow, pH and chlorine but the real challenge is to develop robust multi-variate packages to inform the user on the Astate@ of a water or waste stream. The need is to understand the organic and inorganic content of particular aqueous bodies or streams and an interesting solution might be to develop surrogate and broad band measurements.

For on-line measurement there are two possible routes for the W&WT industry. The adoption of broad band measurements and surrogates and/or the use of multi-variate analysers. Neither approach is mutually exclusive and both need to be supported by and integrated with proper sample collection and analysis procedures and structures.

A EuroEnviron Brokerage Workshop in Manchester October 1996 identified strong interest for measurements of organic load and biomarkers both of which would give imprecise but extremely helpful information.

Data rich multi-variate process analysers could be a source of data for a wide variety of clients within the business. The data would be used, with other data, to derive different information depending on the different needs of the client. Proper integration of these analysers into an open distributed information system would provide significantly more benefits than treating them as mere data providers.

6. Understand the needs

The user's need for sensors might come from a range of considerations; public relations, cost, regulation, prudence, safety, etc ... Sensor technology is changing at an unprecedented rate and key enabling technologies for water monitoring are appearing from unfamiliar sources. Both old and new sensor technologies are constantly being reviewed.

Successful commercial sensors increasingly depend on the blending of a number of technologies and skills; the problem is how to locate and blend those technologies and skills.

The most difficult problem for the W&WT industry is how to develop an holistic understanding of what the user really needs and then to steer the organisation of multi-skilled technology teams to satisfy those needs. There is little doubt that the largest part of the W&WT market for sensors is in the monitoring of trace materials and the market size and growth will depend on the regulator's diligence. The Hazardous Waste Directive (91/689/EEC) provides an interesting list of substances which must be monitored but it does not provide the measurement limits. It is reasonable to assume that the measurements will be at the trace levels with low ppb as a starting point. The Drinking Water 1993 Report by the Chief Inspector Drinking Water Inspectorate, (Ref 3) lists the determinands being regulated together with the levels required and achieved.

It shows that more than 3.5m determinations were made, in 1993, in the regulation of the drinking water sector of this industry and only 1.3% exceeded PCV. Careful reading gives some insight into the needs of the industry.

The UK Environment Agency now monitors 35,000 kilometres of rivers with 10,000 sampling points; each needing at least 12 samples per year. A number of measurands are needed at each sampling point and all of the measurement formats, to be described later, will be applicable. While this is a very large potential market the EA are investing a considerable amount of effort into understanding the sensor needs to support the proper prosecution of their responsibilities. Some of the thinking is in the public domain and it seems that the EA will concentrate on generating a sound understanding of the current state of the UK and its constituent definable areas. This macro approach inevitably means that they will be concentrating on broader measurement concepts and leave the detailed measurement to those who generate pollution. There are consequent influences on the type and number of measurements to be made.
The Water Companies are a ready focus for attention but are only part of the picture. The Regulators, are also influential and are potential users of sensors, but the new Environmental Agency has to settle down and decide how it will operate; this makes planning and forward thinking difficult for the suppliers. A trend to on-site waste treatment and the minimisation of effluent reinforces Industry as an important market.

No reliable statistics are available but an intuitive guess of the split in monitoring device sales suggests :

Water Companies	40%
NRA/HMIP/Factory Inspectorate	10%
Industry	60%

Industry represents perhaps 60% of the environmental sensor market from both the process measurement and control and the monitoring needs and may well, in the longer run, be the most attractive market. Industry includes the such diverse sectors as hydrocarbon processing, pharmaceutical, food, beverage, mineral processing, pulp and paper etc; each with specific problems and sensor needs.

The potable market is almost completely confined to monitoring with almost no closed loop control needs, whereas both the dirty water and the sewage sectors often have control loop implications. The use of closed loop control is expected to increase and it is suggested that suppliers to the dirty water and sewage would enhance their potential if they included process control skills in their offering.

7. Measurement opportunities

It is not difficult to find an opportunity for a new measurement but it is very difficult to present a solution to the W&WT industry in a mutually profitable manner. For example, the trade press has published articles for the prospects for bio-sensors in the W&WT for some years industry but as yet commercial success is extremely limited. One Water plc is seeking better solutions for the following list of measurements:

Aluminium	Clean water
Ammonia	Clean water/Dirty water/Sludge liquors
Bacteria	Raw water
Blanket level	Sludge
BOD(5)	Dirty water/Sludge liquors
Boron	Raw water
Bromate	Clean water
Chlorinated solvents	Raw water
Chlorine	Clean water
Chlorophyl A	Raw water
Coagulant iron	Clean water/Dirty water
COD/TOC	Dirty water
Colour	Clean water/Dirty water
Density	Sludge
Dissolved oxygen	Dirty water/Clean water/activated sludge
Fluoride	Clean water
Hypochlorite conc	Clean water
Iron	Clean water
Manganese	Clean water
Nitrate	Clean water/Dirty water
Ozone	Air
Ozone	Clean water
Particle size up to 0.5mm	
Particle size C	lean water/Raw water/activated sludge
Pesticides	Clean water
pH	Dirty water/Sludge
Phosphate (Total)	Clean water/Dirty water
Polymer concentration	Clean water/Dirty water
Redox	Dirty water
Respiration rate	Dirty water

Sludge	Dry solids content
Surfactants	Raw water
Suspended solids	Clean water/Dirty water
Taste & odour	Clean water
Toxicity	Dirty water
Turbidity	Clean water/Dirty water
Weight	

There does not seem to be a consensus on the detection levels for the vast majority of water quality measurements. A number of commercial companies publish lists but here are significant variations between them and even between different issues of the same document. Anglian Water plc in the UK have put the following list of measurands and limits of detection into the public domain.

Sourceworks Output Quality

	Action Limits and PCV's	
Parameter	*PCV*	*Action Limit*
Colour mg/lPt/Co scale	10	20
Turbidity F.T.U.	1	4
Odour 25C Dilution no.	2	3
Taste 25C Dilution no.	2	3
Odour 60C Dilution no.	5	NA
Temperature C	25	25
pH	6.5-8.5	5.5-9.5
Conductivity uS/cm at 20C	1350	1500
Chloride**mg Cl/l	360	400
Sulphate mg SO_4/l	225	250
Calcium mg Ca/l	low 70 high 160	250
Magnesium mg Mg/l	45	50
Sodium mg Na/l	150	150(80%)
Potassium mg K/l	10	12
Aluminium ug Al/l	50	200
Total hardness mg $CaCO_3$	low 160 high 400	min 150
Dry residues mg/l		1500
Dissolved oxygen (S) % saturation* min 50	-	
Dissolved oxygen (U) % saturation*min 30	-	
Nitrate mg NO_3/l	45	50
Nitrite mg NO_2/l (NH_2Cl)	0.02	0.1
(Cl_2)	0.05	0.1
Ammonium mg NH_4/l(NH_2Cl)	0.25	0.50
(Cl_2)	0.05	0.50
Kjeldahl nitrogen mg/l		1

	Action Limits and PCV's	
Parameter	*PCV*	*Action Limit*
Oxidisability mg0_2/l	4	5
Hydrocarbons ug/l	ND	10
Phenols ug C_6H_5OH/l	0.4	0.5
Boron ug B/l	1500	2000
Surfactants ug/l as lauryl sulphate	150	200
Trihalomethanes ug/l	100	100++
Trichloroethylene ug/l	30	30+
Tetrachloroethylene ug/l	10	10+
CCl_4ug/l	3	3+
Iron ug Fe/l	50	200
Manganese ug Mn/l	20	50
Copper ug Cu/l	1000	3000
Zinc ug Zn/l	2000	5000
Phosphorus ug P/l	high 1500 low 600+++	2200
Fluoride ug F/l	high 1300 low 900+++	1500
Barium ug Ba/l	800	1000
Silver ug Ag/l	5	10
Arsenic ug As/l	25	50
Cadmium ug Cd/l	2	5
Cyanide ug CN/l	20	50
Chromium ug Cr/l	20	50
Mercury ug Hg/l	0.5	1
Nickel ug Ni/l	20	50
Lead ug Pb/l	20	50
Antimony ug Sb/l	5	10
Selenium ug Se/l	5	10
Pesticides ug/l individual	0.05	0.1
total	0.25	0.5
PAH ug/l	0.1	0.2
Benzo 3,4 pyrene ng/l	4	10
Coliforms no/100ml	0	0(95%)
E Coli no/100 ml	0	0
Faecal Streptococci no/100 ml	0	0
Clostridia no/20 ml	<1	<1
Aeromonas hydrophila no/100 ml	0	
Colonies per ml 1 day @ 370E C	10	
Colonies per ml 3 days @ 22E C	100	
Chloride/Alkalinity ratio	1.0	
Alkalinity mg/l $CaCO_3$**	min 55	min 50

++ 3 monthly average +++ where dosed + annual average
*S = Surface Waters *U = Underground Waters **See also Cl/Alk ratio

The UK National Rivers Authority, now part of the Environment Agency have published the following detection limits :

	% bias	% precision	Limit of detection
ammonia (total)	10	10	0.03mg/l
pH	10	10	0.05
dissolved oxygen	10	10	0.2mg/l
TOC	10	10	0.2mg/l
BOD	10	10	1.0mg/l
phosphate	10	10	20:g/l
nitrate	10	10	0.2ng/l
COD	10	10	12.0ng/l
arsenic	10	10	5.0:g/l
cadmium	10	10	0.01:g/l
mercury	10	10	0.01:g/l
lead	10	10	0.1:g/l
copper	10	10	0.1:g/l
zinc	10	10	0.5:g/l
PCBs	20	30	0.3ng/l
DDTs	20	30	5.0ng/l
lindane	20	30	0.3ng/l
dieldrine	20	30	1.0:g/l
atrazine	20	30	10.0:g/l

UK/EC Heavy Metal in Water Concentration Limits:

Material	UK Drinking Water Limit	EC Drinking Water Limit	Discharge Limit 1	Discharge Limit 2	Dangerous Substances Listing
Mercury	1ppb	1ppb	5ppb	50ppb	EC Black/UK Red List
Lead	50ppb	40ppb	200ppb	2ppm	EC Grey list
Cadmium	5ppb	5ppb	20ppb	500ppb	EC Black/UK Red List
Copper	3ppm	-----	300ppb	2ppm	EC Grey List
Nickel	50ppb	50ppb	200ppb	3ppm	EC Grey List
Zinc	5ppm	-----	5ppm	-----	EC Grey List
Chromium	50ppb	50ppb	500ppb	3ppm	EC Grey List

A booklet, prepared by Graham Dennett, Commercial Officer, British Consulate-General, Sydney, in 1995, Ref. 4, lists the water quality measurands and detection levels required by the Australian liquid industrial effluent. It also lists the substances targeted for further reductions in New South Wales by the year 2000. It is unlikely that all these measurements will be made outside the laboratory and even in the laboratory the measurements will be infrequent with samples that may not adequately replicate the water being monitored. This leaves the question as to how the water is to be properly monitored for accidents or other incidents.

8. Summary

Although the market for water quality monitoring is small there are many business opportunities. However it is increasingly difficult to satisfy these opportunities by just concentrating on research and development; technology is only an enabling aspect of useful water quality monitoring instrumentation. A commercially viable product is unlikely to be developed by devoting effort exclusively to pure technology subjects such as optics or biochemistry or electrochemistry. Multi-discipline teams, including a continuous market aware input, are needed to integrate technologies for the provision of solutions to user problems. Such teams will help avoiding getting trapped in the minute details of one sensor in one application field with the consequent small market possibilities. These teams, if properly balanced and led, will also consider the different performance characteristics appropriate for different applications and will track changing needs in the market place.

The key drivers are >business needs=, >business dynamics= and >technology dynamics= and the winners will be those who develop an holistic understanding of the market and an appreciation of the user's real needs before developing products for the market.

An understanding of the 'megatrends' and the 'limitations' is very important to the setting up of profitable projects. For example it is extremely unlikely that all the individual polluting substances will be measured to the detection limits and uncertainties required and hence there is a growing concentration on the use of surrogates. Such broad band measurements as biological activity, oxygen demand, toxicity, nutrient load and possibly bulk pollution measurement are being considered and may prove to be a significant market for monitoring instrumentation. However these surrogate measurements will, by definition, have an imprecise correlation with specific chemicals and calibration audit trails will be an intellectual problem. Most sensor research and development is by scientists, trained in the use of rigorous methodologies, and a cultural change in both the researcher and the user will be required user before broad band surrogates are used with confidence.

9. Conclusion.

It is suggested that future market opportunities will require a realistic assessment of the market sector and the supply of an integrated package of measurement and support services. The ability to make a particular measurement will not be sufficient in itself but merely an enabling feature for the provision of a solution to the user's need.
The challenge to scientists and engineers is to provide simple measurement concepts so that society at large can take part in informed debate leading to a rational control of the industrial and leisure activities.

REFERENCES

1. Green Pounds, by Adrian Wilkes, published in Engineering, May 1995, Europe clamps down on eutrophication. Chemical Engineering, October 1990

2. Sensing needs of the Water Industry. Market Study by Process Measurement Technology for the Laboratory of the Government Chemist, Available from the LGC, Teddington, UK, Autumn 1994.

3. Drinking Water, published by the Department of the Environment, available through the HMSO, 1993.

4. Opportunities in the Australian Liquid Industrial Effluent Sector, Overseas Trade Services, Foreign & Commonwealth Office , London.

Figure 1 : Sensor needs for the water industry

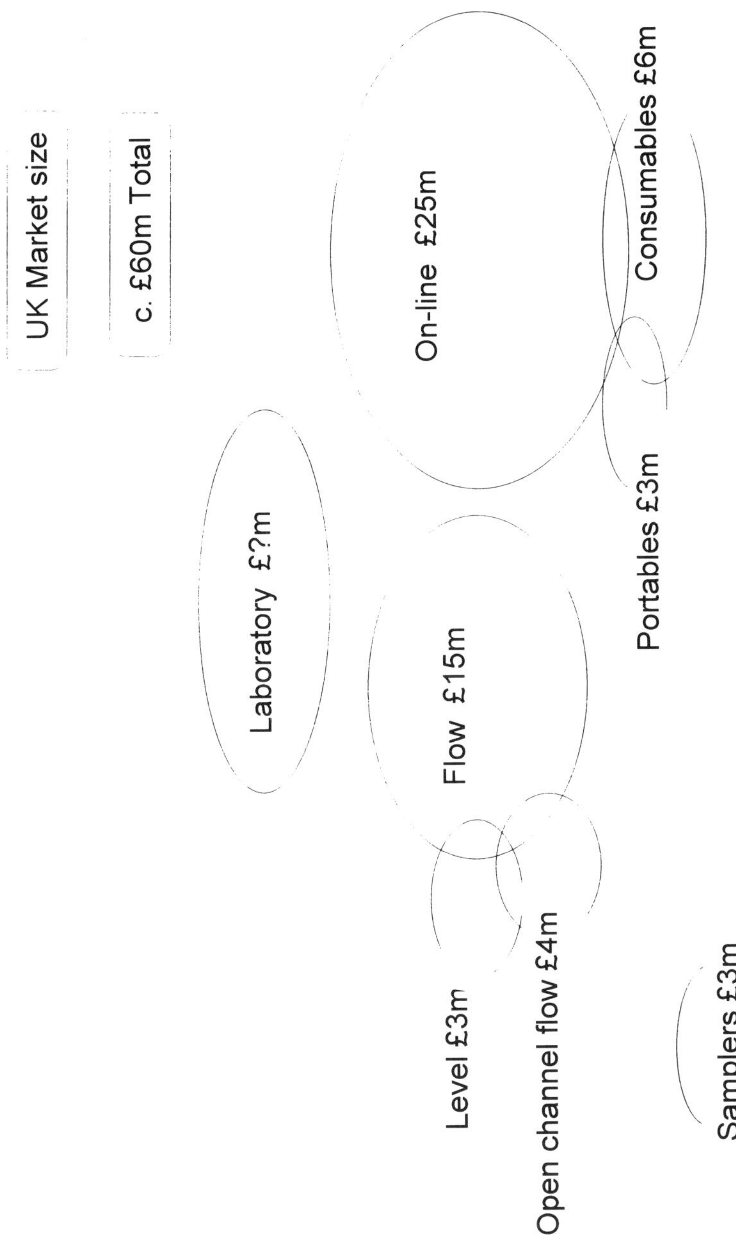

DRINKING WATER QUALITY CONTROL IDENTIFICATION OF NEEDS FOR FUTURE RESEARCH- EPAL'S PERSPECTIVE

M.J. BENOLIEL, Empresa Portuguesa das Águas Livres, S.A., Divisão de Laboratórios, R. do Alviela 12, 1170 Lisboa, Portugal

1. Abstract

Over 2 million people in the city of Lisbon and 21 municipalities in surrounding areas, are being supplied by EPAL, with drinking water produced from surface water and groundwater.

To ensure the safeguard of human health and the compliance of drinking water quality with the Directive's parameter values and national regulation, a systematic monitoring programme is established. To support the water quality strategy, warning stations and on-line sensors are installed, at surface water source and in the supply system.

The paper identifies the EPAL's perspective in what concerns the most relevant needs for research on monitoring of water quality.

Key words: Monitoring, on-line sensors, drinking water quality

2. Introduction

Several Directives dealing with drinking water quality (75/440/EEC[1], 76/869/EEC[2], 80/778/EEC[3]) have been issued in recent years in order to safeguard the consumer. These Directives determine physical, chemical, organoleptic and microbial parameters, as well as their sampling frequency, reference methods of analysis and Maximum Admissible Concentrations (MAC). Drinking Water Directive (80/778/EEC) sets standards for 62 physical, chemical and microbial parameters. An European Commission proposal is currently under discussion within the Member States with a view to revising Directive 80/778/EEC and to including nine new parameters to be analysed. These parameters are carcinogenic and may be present in water intended for human consumption. Their origin may be the raw water, treatment by oxidation (chlorine, ozone, etc.) or treatment products.

Other international bodies, also concerned with drinking water quality, such as the World Health Organisation (WHO)[4] and the US Environmental Protection Agency (EPA)[5], establish Maximum Concentration Levels or recommend Guidelines Values. Both these organisations, establish a significantly higher number of individual organic parameter standards than those stipulated in EC Directives: WHO has Guidelines for 77 organic compounds, while EPA establishes Maximum Concentration Level for 60 organic parameters.

The EEC Drinking Water Directive was incorporated into Portuguese Legislation in March 1990 (DL 74/90).

In order to ensure drinking water complies with the Directive's parameter values, a systematic monitoring programme has to be established by the Water Supplier. The number of sampling points and the sampling frequency is established, either according to the population served, or the volume of water produced or the volume of water distributed[2,3]. For this purpose, analytical laboratories need to have analytical methods capable of producing reliable, accurate and precise data, and attaining detection limits in accordance with each parametric value.

Aware that potable water is one of the most important requisites for human life, the Water Suppliers also develop and set up new structures to tackle the evolution of the water quality concept. Water quality modelling[6,7] and "on-line" monitoring[8] are tools frequently used by water suppliers.

Water quality modelling is being used to characterise, for instance, the optimum sampling location, to predict chlorine decay, bacterial re-growth and nitrate distribution, to define zones where particle deposition may occur, etc.

The importance of continuous and automatic monitoring of the water abstracted, treated and distributed by water supply companies is increasing. On-line analysers provide, from the river intake to the tap, a reliable, precise and efficient tool for drinking water production. Alarm stations, placed a few kilometres upstream from the treatment plants, permit detection of any accidental pollution. The information obtained in "real-time" allows the treatment process to be adapted to the quality at intake or to stop the intake.

In this paper it is attempted to describe the most relevant difficulties encountered during water quality monitoring undertaken by Empresa Portuguesa das Águas Livres, S.A.

3. Epal water quality control

3. 1. Production and distribution of drinking water

EPAL-Empresa Portuguesa das Águas Livres S.A., is a wholly state-owned corporation whose business purpose is to supply and distribute drinking water within its concession area. Two million three hundred thousand people, i.e. about a quarter of the population of Mainland Portugal, are currently being supplied by EPAL in the city of Lisbon and 21 municipalities in surrounding areas.
With a 959 X 10^3 m^3/day catchment capacity and an output capacity in the region of 700 X 10^3 m^3/day, EPAL manages and operates three supply subsystems.

EPAL currently draws water from the following sources :

- Surface water, from the Tagus river (Valada) and the Castelo do Bode reservoir- 740 X 10^3 m^3/day
- Groundwater :

- from limestone soils, in the Olhos de Água springs and the Alenquer and Ota wells- 55 X 10^3 m^3/day;
- from alluvium in the river Tagus, in boreholes in the Valada, Azambuja and Carregado areas- 160 X 10^3 m^3/day;
- from Miocene layers, in the Valada and Lezírias boreholes

Two treatment plants located at Asseiceira and Vale da Pedra treat water drawn from the surface water sources.

Asseiceira water treatment plant includes pre-chlorination, correction of aggressivity by addition of diluted lime and carbon dioxide, chemical coagulation with liquid aluminium sulphate, rapid gravity filtration in a double layer of anthracite and sand and final disinfection with gaseous chlorine.

Vale da Pedra water treatment plant includes pre-chlorination, chemical coagulation with liquid aluminium sulphate, flocculation and settling, using polyelectrolic, rapid gravity filtration in a layer of sand, adsorption on activated carbon and final disinfection with gaseous chlorine.

The distribution network in the City of Lisbon is divided into five zones bounded by distinct topographical levels. It has a length of approximately 1 450 km and consists of 50 to 1 250 mm cast iron, ductile cast iron, reinforced concrete, fibro-cement, fibreglass and high density polyethylene pipes. There are also 18 storage reservoirs having a total storage capacity of 556 X 10^3 m^3.

3. 2. Water quality management

The water abstracted, treated, distributed and delivered by EPAL is regularly and systematically analysed to ensure water quality and compliance with quality parameters defined in Portuguese legislation and by the European Community.
In order to attain these goals, an "off-line" sampling programme is established for the complete supply system, which includes about 220 different sampling points :

- Surface water: 12
- Ground water: 28
- Treatment plant: 13
- Transport and delivery to the municipalities: 102
- Lisbon distribution network: 68

Samples are collected daily for microbial, physical and chemical analysis at the EPAL laboratories.
This periodic check is reinforced by the operation of two alarm units installed upstream of the intake point on the Tagus River. Continuous analysis units are also installed at various points in the supply system. To support its Water Quality strategy, EPAL has the PICCOLO network hydraulic software and chlorine decay modelling in the distribution network is being developed.

Awareness of the risk of contamination derived from the chemicals and materials used in contact with water means that quality checks are carried out regularly to ensure compliance with the agreed specification or CEN standard.

3. 3. Laboratory analysis

At the treatment plants, EPAL has two laboratories, which are responsible for daily water quality monitoring at the intakes and at the treatment plants for process control. Another central laboratory located in Lisbon undertakes all of the analyses in the water quality monitoring programme established within the Supply System, for example at intakes, water treatment, pipelines, points of delivery to the municipalities, reservoirs, etc. These laboratories are prepared to carry out routine analysis of 122 microbial and chemical parameters. They analyse some 20 000 samples each year, requiring around 510 000 individual determinations.

EPAL laboratories are provided with a wealth of sophisticated analytical equipment, which is largely automated and includes :

- High performance liquid and gas chromatography;
- High performance gas chromatography, combined with mass spectrometry;
- Atomic absorption spectroscopy with electrothermal atomisation;
- Segmented flow system (autoanalyser);
- Ion chromatography

The use of electronic sample logging and labelling equipment (optical mark recognition), combined with computerised laboratory information management systems, enables samples to be traced and controlled from the point of reception to the issue of the final report. This system allows rapid and accurate sample processing and reporting of results.

3. 4. "On-line" analysis

3.4.1. The Tagus river receives a large quantity of domestic and industrial discharge, causing a high variability in the quality of the raw water, as well as the occurrence of serious contamination problems. This situation can have a critical effect on the treatment plant located at Vale da Pedra. EPAL has two alarm units, located at the river basin, around 6 Km and 15 Km upstream of the Valada intake. These stations permit the treatment plant to receive, in "real time", information on the following parameters: turbidity, pH, oxygen dissolved, conductivity, temperature, alkalinity, ammonium nitrogen, cyanides, nitrates, chloride, dissolved organic carbon and phosphates.

3.4.2. The surface water intake points are equipped with analysers allowing the control of the treatment process:

- Castelo de Bode reservoir: temperature, turbidity, pH, total hardness, alkalinity and dissolved oxygen.
- Valada - Tagus river: temperature, turbidity, pH, conductivity, dissolved oxygen, Total Organic Carbon and ammonium nitrogen.

3.4.3. Different analysers have been installed for quality control during treatment processes, focused mainly on the monitoring of basic parameters.

- Asseiceira Water Treatment Plant - analysers have been installed in order to control remineralisation and filtration processes and treated water. These analysers permit the detection of pH, conductivity, turbidity, total hardness, aluminium and chlorine;
- Vale da Pedra Water Treatment Plant - analysers have been installed in order to control pre-chlorination, settling and filtration processes and treated water. These analysers permit the detection of temperature, turbidity, pH, dissolved oxygen, total organic carbon, aluminium and chlorine

3.4.4. A network of 60 residual chlorine analysers has been installed along the transport pipelines, points of delivery to the municipalities and Lisbon distribution system. Taking into account the fact that the distribution system is fed with water from different sources, analysers for the control of turbidity and conductivity have been installed at a few selected points in the distribution network.

4. Identification of research needed for water quality management

Daily distribution system management involves coping with changes in water quality in the customer's tap, associated either with the raw water, the treatment plant or the distribution network. In fact, water passing through the distribution system is subject to a number of processes which may affect its chemical and microbial composition :

- Degradation and leaching reactions at the pipe surface
- Changes in chemical species present
- Precipitation and biofilm formation
- Resuspension and film interaction
- Corrosion and degradation in distribution
- Disinfection by-products formation

Considerable attention is being given to these processes. However, the present approach followed by EPAL is based on comparing measurements of the quality in the distribution system with defined criteria and standards. This approach is centred on consumer safety, taking into account parameters that can or cannot be detected directly by the human senses :

- Taste and odours
- Bacterial growth
- Chlorine decay
- Disinfectants-by-products (DPBs)
- Toxic micropollutants

Organoleptic and aesthetic properties of water are the main elements immediately evident to the consumer on appraising water quality. Taste and odours can be produced during the disinfection processes, using namely chlorine, ozone, chloramines or dioxide chlorine. The molecules responsible for the organoleptic properties of water can be mineral elements (metals, dissolved salts) or organics. Some of these organic compounds are the result of biotransformation, while others ones are the result of chemical reactions. At the moment the identification and quantification of odour and flavour causing compounds in water call for a systematic and efficient combination of chemical and sensory methods[9]. Two different procedures for sensory analysis are used: the threshold Odour Number Test (TON) and the Flavour Profile Analysis Method (FPA). However, these tests need to be supported by very expensive identification techniques such as GC-MS and MS-MS. For this purpose it is important to select sensorial standard methods that may be easily applied to water quality control.

Much attention is also focused on the management of the free chlorine present in the distribution network, taking into account the risks represented by microbial re-growth while minimising the risks of taste and odours and the formation of disinfection-by-products (DBP). The efficacy of chlorination is assessed by periodically measuring the free and combined chlorine, using either manual photometer or "on-line" analysis. However, in order to have a better knowledge of chlorine decay within the network and to contribute to the validation of the chlorine quality model, it is necessary to have, amperometric micro-sensors available in the market, at a low cost, easy to maintain and to install, with low detection limit and reproducibility.

Bacteriological monitoring is essential for hygienic control of the water. This is carried out by test methods which only provide an answer 24 hours after sampling the water. If water is contaminated with bacteria, this is far too long to wait without reacting with the requisite water treatment and warnings to the public. Accordingly, a rapid screening technique is necessary to assess the presence/absence of bacteria and/or enumeration of coliform bacteria[10]. To this aim, instruments need to be developed either for laboratory or "on-line" monitoring.

As mentioned above, a systematic monitoring of drinking water quality has to be designed for each individual "supply zone", according to EC Directives and national regulation. For this purpose Water Companies need to have, within their bodies, analytical laboratories able to produce reliable accurate and precise data. Moreover, results have to be quickly available, in order to allow, for adequate corrections within the distribution system as and when necessary.

Analytical laboratories have nowadays access to techniques for inorganic parameters, which have proved to give reliable and accurate results, such as Ion Selective Methods, Inductively Coupled Plasma (ICP), Ion Chromatography (IC), Atomic Absorption Spectrometer (AAS), etc. The problems arise when the detection limit is too close to the Guide Value, the measurements being made in a concentration range of great variability and uncertainty (for example, lead and arsenic). Research should be strengthened for the development and validation of the multi-element analysis, namely by atomic absorption spectrometry, improving their sensitivity. The development of "on-line" sample pre-treatment with final detection technique is also necessary.

During the last decade legal requirements have become increasingly demanding in what concerns the presence of organic compounds in drinking water. The problems of analysing organic micro-pollutants in water relate mainly to their large variety and very low concentration levels.

Analyses of organic compounds are generally performed with sophisticated equipment, need knowledgeable operators, are very expensive and the results are not quickly available.

There are currently several potential "priority" organic compounds, namely pesticides, for which there is no available method capable of monitoring samples at the limit level of 0.1 μg/L. In addition, many methods are inadequately tested at this concentration and are not accurate enough for the purpose of monitoring compliance. Systems should be developed and made available commercially, regarding "on-line" extraction and clean-up with final detection.

Regarding "on-line" analysers, the sensors commercially available are dealing mainly with inorganic compounds like anions (nitrate, nitrite, phosphate, cyanide, ammonium) and heavy metals (lead, copper, zinc). For organic compounds, it is necessary to develop sensors with sensitivity and specificity. As a rule, it is recommended to put more effort into developing and validating "on-line" analysis for the concentration level of standards and guidelines. For these analysers, it is also essential to guarantee a representative and constant flow of the sample water in order to obtain accurate and reliable results.

5. Conclusions

Several areas where research should be increased are identified :

- Rapid screening techniques to assess the presence/absence of bacteria and/or enumeration of coliform bacteria, for laboratorial and "on-line" purposes;
- Development of sensorial standard methods for taste and odour, which may be easily applied in water quality control;
- Chlorine micro-sensors to install in the distribution network;
- "On-line" analysers for disinfecting by-products (THM's, bromate)
- Fast toxicity tests for drinking water purposes;
- The research needed on the analysis of organic micro-pollutants can be summarised:
    - Lower detection limits
    - More accurate analysis at low concentration values
    - Development of low cost and rapid methods
    - Development of fully automated analysis (sample pre-treatment and final detection technique)
    - Development of sensors with sensitivity and specificity
- Research on inorganic analysis:
    - Development and validation of the multi-element analysis, improving their sensitivity.
    - Development of "on-line" sample pre-treatment with final detection technique is also necessary
- Development of fast methods for assessing biological stability:
    - Assimilable organic carbon (AOC)
    - Biodegradable Organic Carbon (BDOC)
- Development of reliable test procedures to assess effects of distribution materials on water quality

6. Acknowledgement

The author wishes to thank all the colleagues who provided information for this paper.

REFERENCES

1. Official Journal of the European Community, L 194, pp 26, Council Directive 75/440/EEC "Surface water for drinking", 25/7/1975.
2. Official Journal of the European Community, L 271, pp 44, Council Directive (79/869/EEC) "Drinking water measuring methods and sampling frequencies", 29/10/1979.
3. Official Journal of the European Community, L 229, pp 43, Council Directive (80/778/EEC) "Water for human consumption", 30/8/1980.
4. WHO. "Guidelines for drinking-water quality", Vol. 1. Recommendations, Geneva, 1993.
5. F. W. Pontius- "Future Directions in water quality regulations", Journal AWWA, Vol. 89, Issue 3, pp 40, 1997
6. M. R. Gwynne and M. F. Williams: in IWSA specialised conference on" New developments in modelling, monitoring and control of water supply systems", pp, 29, 24-25 September 1996.
7. P. Piriou, Y. Levi, J. Heraud and L. Kiene. in IWSA specialised conference on" New developments in modelling, monitoring and control of water supply systems", pp 137, 24-25 September 1996.
8. G. Fr. Asdgaard, J.D. Berg, B. Nesgard, H. Ratnaweera, B.M. Wathne, K. Dagestag and G. Lenes: In IWSA 20th International Water Supply Congress and Exhibition on "Water supply - The essential Service: working for Excellence", SS 13, pp1-4, Durban 9-15 September 1995.
9. Joel Mallevialle and I.H. Suffet- "Identification and treatment of tastes and odours in drinking water"- AWWA/Lyonnaise des Eaux, 1987
10. Th.H.M. Noij, A.Brandt &M.M.E. van der Kooi: In IWSA 20th International Water Supply Congress and Exhibition on "Water supply - The essential Service: working for Excellence", SS 13, pp5-8, Durban 9-15 September 1995.

SENSORS FOR WASTE WATER : MANY NEEDS BUT FINANCIAL AND TECHNICAL LIMITATIONS

H. WACHEUX, Anjou Recherche, B.P., 76, 78600 Maisons-Laffitte, France

1. Introduction

A waste water system can be divided into three steps, from upstream to downstream, according to different responsibilities of operation :

- discharge of waste water in the sewer, especially industrial discharge
- collecting network (combined or separate) and combined sewer overflows
- waste water treatment plant and discharge of epurated water in receiving water.

We will examine separately the needs at the above three levels as well as the current practice, because they are very different when considering them technically, financially or legally.

Most of the online measurements are performed after treatment or pre-treatment in order to check compliance with a predefined consent level. Recent environmental directives in EEC are pushing in parallel the various ways of measuring water quality : laboratory analysis, field instruments and online monitors, especially at discharge level.

However online monitoring is now also expanding in control system for the automatic or man assisted operation of the water works.

2. Present level of equipment

If we let aside flowrate measurement and level measurement, online monitors are still rare in waste water systems, especially in collecting sewers. This is in relationship with the size of the collecting networks, and with the fact that this subterranean part of the system is ignored by most of the people.

For the various agents in water field - private or public operators, communities, river authorities - it is different. Due to historical reasons, and because everyone is reluctant to change his own habits, control and operation of waste water systems are mainly based on sampling and laboratory analysis.

Figure 1 : analysers parts

ELECTRONICS	Traditionally, an automatic analyser can be divided into three parts :
HYDRAULICS	The with electronic components are located at the top, a screen and a keyboard.
REAGENTS STORAGE	The pumps and piping are in the middle part, generally behind a glass door The lower compartment is for reagents storage.

Online monitors are rarely considered, though they show nice features such as :
response time : a few seconds or a few minutes instead of a few days.
time covering : the overall yearly period is covered, which is not ordinary the case with daily samples.
automation : from sampling to result presentation, every step is automatic. This allow for automatic reaction of the system to adjust to new conditions or automatic warning of the human agent.

At the level of a discharge into the public network, main industrial sites or most pollutant ones are progressively obliged to install online monitoring of water quality. They often start with flowrate measurement and automatic sampling, then temperature, pH, conductivity, O2 or rH. In some case they follow with TOC monitor, which is much more expensive. According to our information, about 500 TOC monitors are installed in France on industrial discharges in order to follow up organic pollution load.

Figure 2: automatic analyser for ammonia using gas electrode

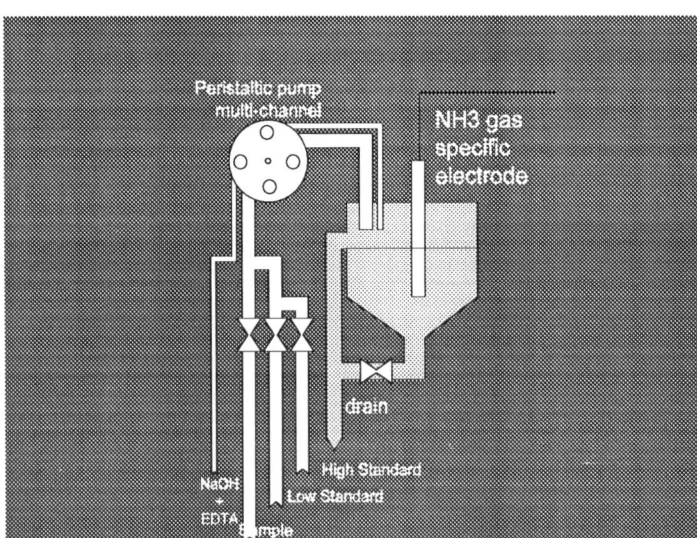

In collecting network, online instrumentation is very poor, in almost every case it concerns level or flowrate measurement. The reason is that flow is made by gravity, there is no need and no way to adjust the flow and for the same reason it is not necessary to measure it except in pumping stations. But things are changing, because of the public concern about Environment and water quality.
Public authorities are imposing tougher and tougher checking especially at overflows, which are very numerous and responsible for a large part of pollution during rainfall events in combined systems. Continuous monitoring of collecting network is encouraged, level detectors are installed in overflows, in some cases they are completed by flowrate measurements, automatic sampling or quality measurements such as turbidity, dissolved oxygen, ect ...

Online instrumentation is not yet common at the inlet or the discharge of waste water treatment plants. A recent enquiry from AGHTM (the French Professional Water association) and not yet published, has shown that online sensors are almost non existent in present plants under 50 000 PE and represent only a few percent in bigger ones.

For process control, inside the plant, online sensors are less rare, about 4 percent of the plant have at least one sensor. If we except flowrate or level measurements which are rather common, in the water file we could find mainly sensors for dissolved oxygen or O.R.P. in activated sludge. In some occasion they are completed by TSS at the inlet, after primary settling tank, in activated sludge or return sludge or return mixed liquors, and sludge blanket level in settling tanks. In the sludge file we could find sludge blanket level in thickeners, TSS at various steps, pH and temperature in digestors, sensors related to air quality and staff security (CH4, NH4, explosion index, etc...).

Figure 3 : waste water treatment plant, low rate activated sludge

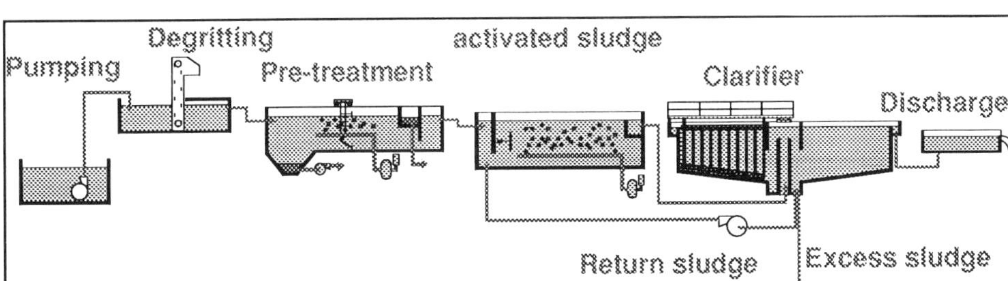

3. Regulation as a support to online monitoring.

In France, industrial wastes are driven by the ministerial order dated 01 March 1993, to be modified in a next future. This order gives a limit value for every parameter related to organic pollution (COD, BOD, TSS ...etc.) but also for various other pollutants more specific of industrial activity. It is well known that some organic or mineral by-products of industrial activity are non biodegradable or have a toxic effect, sometime very far from the emission point.

This is the case of heavy metals in particular, which are concentrated in the activated sludge of waste water treatment plants or in the sediments of the rivers or along the biological chain in the ocean.

The French ministerial order dated 22 December 1994 is related to monitoring waste water systems. It uses the parameters defined in EEC directive dated 21 May 1991, that is to say COD, BOD, TSS and over that total Nitrogen and total Phosphorus for sensitive areas. The frequency of analysis for every parameter is defined according to the size of the system : it varies from 4 to 365 analysis of daily samples per year, that is to say once a day.
The means of measurement are not specified, by default it is laboratory analysis, so to substitute it by online monitoring a preliminary agreement must be found with the various administrations in charge of water quality control.

The alternative of continuous monitoring is suffering from a lack of credibility in public institutions. This comes from the fact that a new parameter always comes with a standard referring to laboratory analysis, in order to define how to measure it. Continuous measurement comes later on, and naturally it differs slightly ore deeply from this standard protocol. Therefore online measurement is suspected to lead to a second choice measurement compared with laboratory analysis.

For instance a satisfactory operation percentage of 60 % of the time is often met for an automatic analyser, which is disappointing for the end user or not admissible for a control loop. A high educational level , intensive maintenance and cleaning are often required, and this is not always adequate with the existing staff which is not as specialised and numerous as in an industrial plant.
There are not enough standards and specifications to guarantee a predefined level of quality for online monitoring, so that both the user and the administration can rely on online monitoring. A project named ETACS, supported by EEC commission DG XII, has the objective to produce a pre-normative standard protocol for the evaluation of waste water quality analysers.

4. Technical aspects of continuous measurement.

There are many ways to associate an instrument and a process line : the instrument can be in line, on line, at line, off line. The instrument itself is not alone, it is a link in a measuring chain which has many other links :

- sampling and transfer of the sample,
- division of the sample, a fraction is taken and transferred again,
- conditioning of the sample before analysis
- analysis of the sample, in a single step or after several steps
- electrical output, transfer, data is digitised, processed and presented to the user

Referring to industrial waste monitoring, useful parameters are very numerous. The basic ones are TSS, COD, TOC and BOD.

The other ones are very specific of every activity and even every site : heavy metals are common in surface treatment and mechanical industry, chromium in tanneries, oil in petrochemical industry, AOX, PCB in industries using chorine. An other aspect which is a challenge for online monitoring is the complex meaning of toxicity. In some cases, it is very important to appreciate continuously and rapidly the risk of releasing a waste in a biological system located downstream.
This starts often by introducing the waste in the treatment plant itself, because it is very often based itself on biology. On an other side water like other elements follows a cycle which includes many successive uses along the watershed, and a part of it becomes drinking water for human communities.

In collecting networks, needs of online monitoring are concentrated in overflows, and they are similar to those related with discharge into waste water treatment plants. Once again basic parameters are COD, BOD and TSS, with total N and total P in sensitive areas. Technical constraints are stronger in a sewer, the accessibility is bad, the space is limited, and environmental conditions are not favourable to instrumentation (humidity, darkness, flooding, strong variability of flow, detritus of all size in water ...etc.).
The network itself is large, so it cannot be monitored at every node, a hierarchy of the needs must be set. When starting from nothing, which is generally the case, some parameters are already accessible online and are not too costly, such as turbidity, pH, conductivity, dissolved oxygen, O.R.P. and organic load indicators.

At the inlet and at the discharge of a waste water treatment plant, the law has fixed the parameters to follow. Some of them are not accessible to online monitoring, some other are difficult or costly to measure online, but they can be derived from more accessible parameters, for instance turbidity to estimate TSS and COD, UV and visible absorbance to derive COD, NH4 and NO3 to derive total Nitrogen, PO4 to derive total phosphorus.

We still need automatic analysers measuring the imposed parameters :

- total phosphorus instead of PO4
- total nitrogen instead of NH4, NO2 and NO3
- COD instead of pollution indicators
- BOD or a substituted standard parameter
- TSS instead of turbidity

Table 1 : list of needed sensors and analysers in waste water systems

parameters	Exists?	Cost in kF	Comments on needed features
pH	+++	10	ruggedness, auto-cleaning, auto-calibration
rH	+++	10	automatic cleaning, prevention of contamination
O2	+++	15	automatic cleaning and automatic calibration
H_2S	++		analysis in sewers, very harmful
TSS at inlet	0	70	via turbidity, automatic cleaning of detritus, grease
TSS at discharge	0	25	low cost, auto-cleaning, without maintenance

parameters	Exists?	Cost in kF	Comments on needed features
sludge settlability	+		equipment derived from lab, too complex & costly
siccity of sludge	+		field test, not yet online instrument.
Sludge blanket level	+++	30	optical absorb. or U.S. reflection, need of auto cleaning and sensitivity
NH_4^+	+++	80	low cost, reliability, not too much maintenance
NO_3^-	+++	80	low cost, reliability, not too much maintenance
total nitrogen	++		too complex, too costly actually
PO_4^{3-}	+++	100	low cost, reliability, not too much maintenance
total phosphorus	0		not really available online, not effective
COD	++	200	derived from lab, low cost and reliability needed
BOD	++		far from standard (5 days)
TOC	+++	140	low cost, reliability, not too much maintenance
biodegradability	++		concept to be clarified and standardised
toxicity	++	to 400	standard, low cost, light maintenance, interpretation

Keys for Existence : +++ : well developed, ++ : need some " tuning ", + : still in infancy
0 : not yet available

For dewatered sludge, needs are specific : pH, siccity, mass and volume flow, biological stability index, odours, lime content, available fertilisers N and P, micro-pollutants such as heavy metals (Hg, Cr, Cd, Cu, Pb, Ni, Se, Zn), HAP and PCBs.

5. Conclusion

Technically, the needs of online monitoring are strong in order to operate and to follow up the waste water systems, but more than technical aspects, the lack of standards and the high cost are the mains reasons why online monitoring is not yet developed in waste water.

Today, the cost of a cheap sensor such as a pHmeter is less than 10 kilo Francs, and at the opposite, a toxicity analyser using respirometry or bioluminescence is more than 400 kilo Francs. And naturally, the owner's cost is not limited to simply buying the instrument. For instance the cost of an analyser for NH4, NO3 or PO4 is about 150 to 200 kilo Francs, including installation. This high cost of investment is exceeded by the operating and maintenance cost.

Technical progress and advanced technologies tend to lower these prices and therefore allows the use of online monitoring where it was not affordable a few years before. But some other aspects are also problematic, such as reliability and easiness of operation, they can be solved by user's feedback and manufacturing at a larger scale. It is always surprising to see that high cost of an analyser is often joined with lack of reliability and complexity. Training of the operators is one of the keys to success.

An other key is specification, a clear methodology must be followed before obtaining online measurements : Users needs definition, supplier consulting, choice based on financial and technical evaluation, well defined ordering, installation according to standard guidelines, commissioning, adequate maintenance and operation.

REFERENCES

1. Charpentier.J., Godard.H., Martin.G., Mogno.Y." Oxidation-Reduction Potential (ORP) regulation as a way to otpimize aeration and C, N and P removal ", Experimental basis and various full-scale axamples, Wat.Sci. Tech., vol 21, 1209-1223, 1989.
2. Andersen K.L. and Wagner P., " Testing on-line monitors for ammonia, nitrate, orthophosphate and suspended solids in activated sludge sewage treatment plants, ICA proceedings of 5th IAWPRC Workshop, Yokohama and Kyoto, Japan, 26 July - 3 August 1990, p667-675, 1990.
3. Wacheux H. " L'instrumentation dans les stations d'épuration urbaines " Proceedings of Hydrotop 92, Marseille, 8-12 April 1992.
4. Wacheux H., S.Da Silva and J.Lesavre, " Inventory and assessment of automatic nitrate analyserz for urban sewage works ", Proceedings of 6th IAWQ Workshop,Ontario, Canada, June 21-25 1993, p. 529-538, 1993.
5. Wacheux H. (1995) " Super Oxydose, un automate pour les petite usines d'épuration ", Proceedings of technical meeting " Automatique et Dépollution biologique ", INRA, 9 p, 05 December 1995.
6. Wacheux H., F.Rogalla, " Sensible selection of sensitive sensors for sewage nitrogen removal ", proceedings of WEFTEC 96, session 72, vol. 6, p 523-533, 1996.

NEEDS AND USES OF CONTINUOUS MONITORING EQUIPMENT FOR WASTE WATER TREATMENT. STANDARDIZATION Y/N ?

J.L. CECILE, NAN.C.I.E. / G.EM.C.E.A., 149 rue Gabriel Péri F-54500 Vandoeuvre, France

1. Abstract

Users and producers of continuous monitoring equipment (on line, in line or in situ) have expressed the need for standards describing the specifications, performance tests, procedures of use including maintenance and finally certification. But what are the real needs of this standardization, mainly for the treatment plants of waste water. From the European directive 91-211, and the application law in each country in Europe, users, owners of plants or their operators have needed to control the effluent coming into the natural water.

Consequently, there is an urgent need for improvement of comparability, reability and quality of measurements obtained for continuous monitoring equipment used to determine the chemical composition of waste water in order to support the protection of environment.

At the moment the main objective is to dispose of parameters for the control of processes and the measurement of the water quality to apply the regulation.

In term of technical point of view for the all application fields, the problems are approximately the same.

In the future, the users claim to have a regulation application including the indirect and global parameters. An important R and D study is to standardize the principle and the application methods for these parameters.

In conclusion, a standardisation of equipment for continuous measurement and also a standardisation for the procedures to apply them is necessary.

2. Introduction

For a long time, the operators of waste water treatment plants have been looking for a continuous measurements to improve the control of process. Many parameters are able to measure : basically parameter as flow, temperature, pH, Redox, turbidity, ... but also chemical parameters as nutrients : nitrate, ammonium, orthophosphate, ...
Due to the very heterogeneous medium and complex procedure of use the equipment : sensors, analyzers are poorly used with success.

Users and producers of continuous monitoring equipment (on line, in line or in situ) have expressed the need for standards describing the specifications, performance tests, procedures of use including maintenance and finally certification. But what are the real needs of this standardization, mainly for the treatment plants of waste water. From the European directive 91-211,(1) , and the application law in each country in Europe, users, owners of plants or their operators have needed to control the outlet coming into the natural water. Four needs are involved,(2) :

- obligations from regulation
- contract between the " actors " : users, operators, suppliers
- technical solutions : R and D of new equipment
- economical contents : cost of investment and use.

Consequently, there is an urgent need for improvement of comparability, reability and quality of measurements obtained for continuous monitoring equipment used to determine the chemical composition of waste water in order to support the protection of environment.

At the moment the main objective is to dispose of parameters for two fields,(3) :

- control of process
- measurement of quality to apply the regulation.

In term of technical point of view for the all application fields, the problems and solutions are approximately the same. Finally, there are three types of parameters :

- parameters in application of need for process or regulation, named direct parameters
- parameters in substitution, with the direct parameters, there is a perfect correlation between them, named indirect parameters
- parameters as indicators, without correlation with the previous parameters they give an information of variation, the main subtype is named global parameters.

In the future, the users claim to have a regulation application including the indirect and global parameters. An important R and D study is to standardize the principle and the application methods for these parameters.

The main difficult is at the moment to measure with a relatively good efficacy the parameters fixed by the technical application of regulation. In France, the European Directive 91/271 has been translated by the low on January 1992, named " Low for water ". Three parameters in general application are : BOD_5, COD, SS (solids in suspension), for the sensitive eutrophisation areas two other parameters must be measured : TNK (Total Nitrogen Kejdahl) and TP (Total Phosphorus).

Unfortunately, there is any easy method to measure continuously these five parameters. The usual procedure consist in a sampling and a determination in laboratory by standardized methods. The batch measurements by daily samples give buffered information, the result is a mean of actual values, as shown in figure 1. For this first reason solutions are reached to have an information from a continuous measurement, named continuous monitoring.

This term includes all measures obtained in line, on line or in situ, in real time or with a minimal delay (few seconds or minutes) . The procedure is other a true continuous measure or a sequential one.
The true continuous measures are physical phisico-chemical parameters : pressure, temperature, conductivity, pH, Redox, certain selective electro chemical sensor, or optical sensors. The sequential measures are obtained by analyzers consisting in an equipment reproducing in continuous procedures the laboratory methods, there are mainly absorptiometric principles : nitrate, ammonium, orthophosphate, toxic compounds as cyanide, heavy metals. The delay attains 20 to 30 minutes for example for orthophosphate.

The accuracy and resolution of these procedures are very good, but the main problem is the maintenance in term of complexity and cost. In the other hand, the equipment are generally very expensive, the table 1 gives an idea of effective costs by uses.

The use of indirect and global parameters is cheaper in terms of investment and maintenance. It needs to know very well the system where it is used : volume and quality of effluents, risks of variation by rain water and industrial waste water.
This last point is very hard, to obtain an exhaustive knowledge of pollutants coming very industries and the evolution of their effects in waste water effluents.

3. Description

In order to support the protection of environment and to respect the obligations of the new regulation, there is an urgent need for improvement of comparability, reability and quality of measurement obtained from equipment for continuous measurement used to determine the chemical parameters and flow of water in
waste water treatment plants but also intakes of surface water for drinking water production plants, industries effluents. This need is expressed by users and producers, the authorities in charge of regulations application are waiting for an agreement to promulgate official use.

Figure 1 : Variation of NH_4^+ concentration for a final effluent of sewage plant

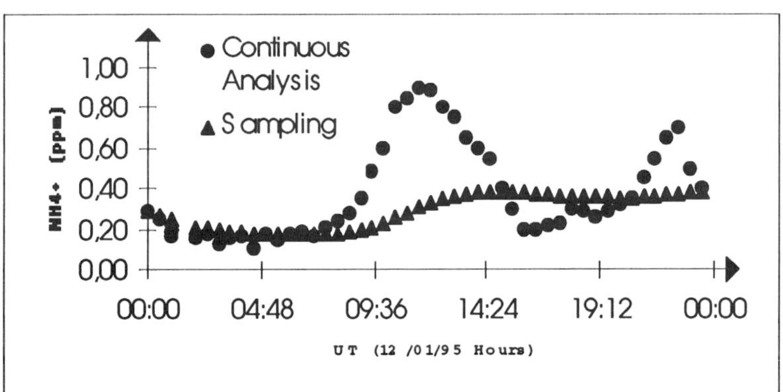

The objectives of this standardization will be :

- the description of technical specification of equipment
- the performance tests giving actual values to the technical specifications of equipment
- the operational procedures, checks, maintenance which will secure reliable results of equipment
- the definition of parameters type for specific application.

In fact, the usual approach consists to apply an equipment to obtain the equivalent result obtained by sample - laboratory analyses method, this is the main error, because it seems at the moment impossible to reach this procedure.

In resume, three groups of parameters are considered :

- in agreement with regulation BOD_5, COD, SS, TN, TP
- in substitution with parameters in agreement with regulation
- as indicators of pollution.

For the first group, there is any easy and simple equipment, 10 obtain correctly and rapidly values to be used either for regulation control or for process control. The investment and maintenance cost are very high. The use requires high technician person. In the second group, the parameters are obtained by simple equipment but the measures are correlated with results of usual procedure for previous parameters. There are a lot of example for this approach. But the correlation are not obtained in a controlled procedure. For example, the main cases are :

- TOC in substitution of COD
- Turbidity in substitution of SS
- UV absorption (254 mm) in substitution of COD.

Table 1 : Estimation of Costs for Continuous Measurements (from AGHTM's report to be published)

Parameters for regulation

	Units	DCO	COT	Ptot	Ntot
1- Investment					
Furniture Sensors :range of price	kF	180-250	100-160	100-180	100-160
Furniture Sensors : mean price	kF	200	130	150	130
Sampling, conditioning	kF	20	20	20	20
Installation	kF	25	25	25	25
Check in of chain of measure	kF	p.m.	p.m.	p.m.	p.m.
Duration of investment	year	7	7	7	7
Sub total of investment	kF/year	32	22	25	22
2- Cost of use and maintenance	Daily Cost for one man : 2 kF/day				
Duration of maintenance (one man)	j/year	27	24	24	24
Sub total Manpower	kF/year	54	48	48	48
Reagents + pieces in replacement	kF/year	30	20	20	20
Validation periodic of the chain	kF/year	p.m.	p.m.	p.m.	p.m.
Sub total for Use and Maintenance	kF/year	84	68	68	68
TOTAL.....	kF/year	116	90	93	90

Parameters in substitution

	Respir	Turbidity	Turbidity	PO4	NH4	NO3
1- Investment						
Furniture Sensors :range of price	250-400	40-80	20-40	80-120	70-120	60-110
Furniture Sensors : mean price	320	65	30	100	95	85
Sampling, conditioning	0	0	0	20	20	5
Installation	30	5	5	1	25	5
Check in of chain of measure	p.m.	p.m.	p.m.	p.m.	p.m.	p.m.
Duration of investment	7	7	7	7	7	7
Sub total of investment	50	10	5	14	17	13
2- Cost of use and maintenance						
Duration of maintenance (one man)	48	6	3	12	12	6
Sub total Manpower	96	12	6	24	24	12
Reagents + pieces in replacement	12	6	2	20	12	6
Validation periodic of the chain	p.m.	p.m.	p.m.	p.m.	p.m.	p.m.
Sub total for Use and Maintenance	108	18	8	44	36	18
TOTAL.....	158	28	13	58	53	31

Figure 2 : Linear regression for parameters in substitution and parameters for regulation

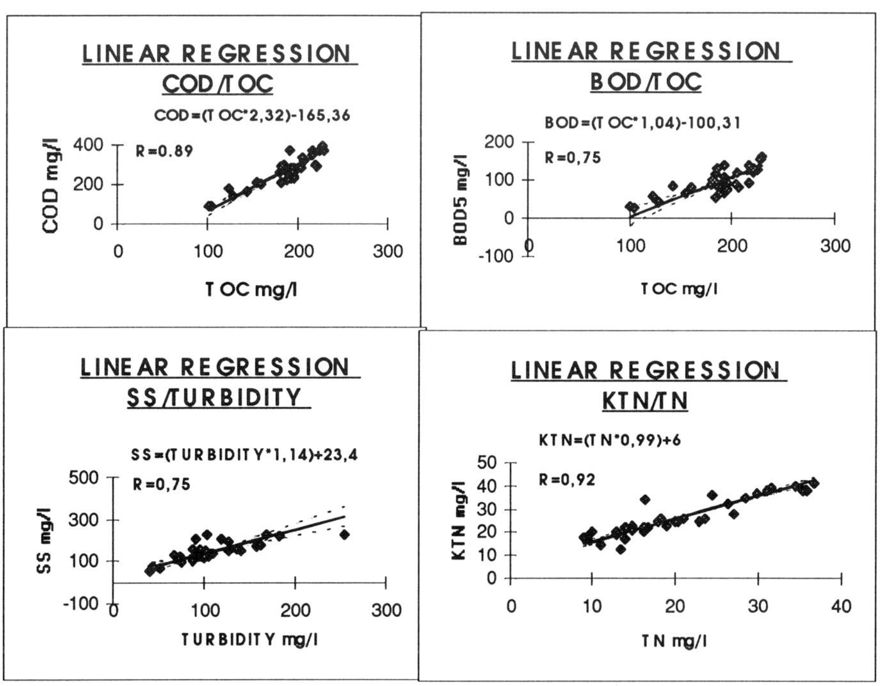

In an homogeneous medium, with constant quality in terms of matter in suspension and dissolved, the correlation given by a linear relationship between a direct parameter and the parameter in substitution is good, fig.2. But in actual applications, these results are often no true. This approach requires a measurement of parameters in interference with the main parameter, consequently it needs to have a multiparameter measurement.

The parameters as indicators are obtained by very simple equipment, it is necessary to correlate result with the presence and the variation of regularly parameters and also with environmental parameter of measurement. This group of parameters includes global parameters :

- pH, Redox, Conductivity
- Solids in suspension
- Total Organic Load
- Biodegradable Organic Matter
- Toxicity
- Odors

The effective measurement are obtained by physico chemical parameters : electro chemical, optical, but also by respirometry or biological system, using bacteria, algae's, fishes,...

An example is given fig.3, the problem was to control influent of an industrial waste water coming from metallurgical plants using acid baths contaming heavy metals (Pb, Cu, Zn, Cr...). It was impossible to measure the precise concentration of these metals compounds, finally a multiparameter probe has been used including pH, T, Conductivity and Redox. The combination of parameters has given in line earlier and sensitively an indication of small quantity of pollutant influent. The system has included a sampler to make precise analysis in laboratory. In the future, the waste water effluent could be control to prevent the risks for process in sewage plants.

Figure 3 : Effect of polluants from industrial effluent on domestic waste water quality

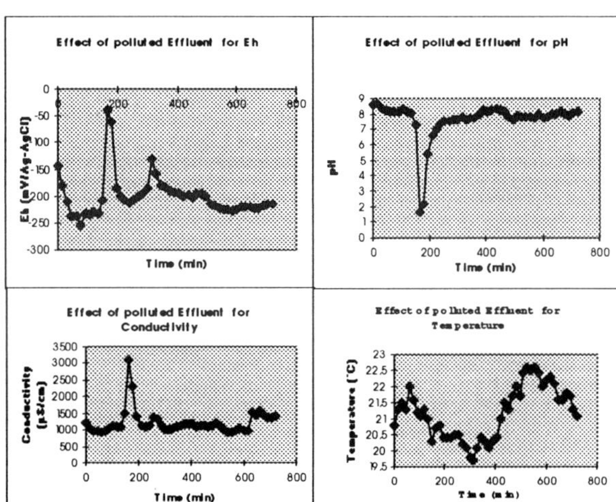

Many groups of research are working with this problems, for example in US there is an ASTM'S designation : Standard guide for continual on line, Monitoring Systems for water analysis. Recently Danish Standard Institute has proposed to create an ISO's working group named on line in situ sensors/analyzers for water. This approach is coming up from a research group ETACS, including Danish, French, English and Spanish researchers and suppliers. Specifications, performance tests and possible certification procedures.
Another group is working in CEN, CENELEC (BT WG 70-3) with European standards for process instruments, the leadership of this group is WRC in UK.
A group has been found in France at the beginning of 1997 to propose a French position. this group is an association of users, producers and authorities, users are owner of plants and operators.

In G.E.M.C.E.A., there are in development facilities as pilot plants to improve and apply procedures of evaluation :

- hydraulic channel for flow-meters, shown on the figure 4,
- rapid closed circuit for samplers and turbidimeters,
- pilot plant for sensors and analyzers of specific pollutants.

Figure 4 : Picture of hydraulic channel to test and to evaluate equipments in GEMCEA

4. Conclusion

Users and producers, but also authorities have expressed the need for standards of continuous monitoring equipment applied in water quality.

For this is necessary to define the parameters in good agreement with application, after that technical specifications are needed to be described. The performance tests are propose to confirm adaptation of equipment with uses. And finally all operational procedures must be perfectly given to reach good results for a better control.
In conclusion, YES, standardization of equipment for continuous measurement, but also standardization of procedures of use, is necessary :

- to control the quality of effluents in agreement with regulation,
- to control end improve the processes of treatement,
- to reference the specifications in contracts between for example users and suppliers.

REFERENCES

1. BLÖCH H., Urban waste water treatment in European Union, Water Management International, p.7-10, 1997.

2. JOURNET J.M., LAVARDE R., CECILE J.L., VERDEAUX P., Mesures en continu et autocontrole des systèmes d'assainissement : de l'usage à la normalisation, 76ème Congrès de l'AGHTM, Londres, 28-31 mai, p.49-68, 1996.

3. VILLESOT D., FUSILLER P., HYVRARD N., STIVANI P., GEORGES T., Mise en place de l'autosurveillance des stations d'épuration : une contribution à la qualité dans les services aux collectivités, TSM, n°9, p. 605-9, 1996.

IMPROVEMENT OF WATER QUALITY SURVEILLANCE WITH THE DEVELOPMENT OF A NEW APPROACH: INTEGRATED CONTROL OF PHYSICO-CHEMICAL AND BIOLOGICAL SENSORS RESPONSES

J-P. SCHMITT and. M. THOMAS, NANCIE : Centre International de l'Eau, 149, rue Gabirel Péri, Vandoeuvre, France

1. Definition of the problem area

Water quality resources must be closely supervised to ensure that they respond fully to safety requirements associated with its different uses. Yet water in general is constantly subject to ever increasing risks of pollution, concurently more stringent quality requirements are imposed on water for human consumption for example. Several thousand xenobiotic substances may be expected to be found in a river although most of them are in very low concentration and sometimes undetectable through routine analytical methods. It is essential to provide the scientific and technical bases for improved water surveillance capable of global and rapid detection of pollutants from accidents or malevolences in order to protect surface waters (aquatic life, drinking water resources).

The cost of monitoring programmes is really low compare to the economic and social impacts of pollution which are not detected early enough to undertake adequate action.

2. Current solutions and limitations

Continuous surveillance of river water quality has recently attracted growing attention. Water and effluent quality measurement is routinely based on the analysis of random samples. Spot samples are collected periodically and transported for laboratory analysis. Not only is the evaluation of the results problematic for the periods in between the sampling events, but due to financial and analytical constraints, only a limited number of compounds can be detected chemically. Although continuous monitoring is possible for some chemical parameters, it is widely accepted that routinely used chemical monitoring and analysis methods only detect a limited fraction of the toxic compounds that may be present. In addition, physico-chemical surveillance does not take into account interaction phenomena.

To cope with this situation, a new solution relies on the use of biological sensors allowing a wide monitoring spectrum. These biological systems indicate that a harmful condition exists, even when the compound cannot be detected analytically by routine monitoring programmes. Each organism has its specific sensivity towards certain pollutants but also towards environmental conditions. Since changing natural conditions may affect the parameters measured, knowledge of such effects is needed to improve data analysis and evaluation.

On the other hand, there is a wide range of reported lethal and sublethal concentrations for different organisms deducted from laboratory tests, yet there is much confusion as to how this knowledge should be integrated into water quality research. When comparing various investigations on the performance of biological early warning systems and biosensors, differences in the outcome have frequently been observed.

In relation to this problem, this research project aims to develop a multi-system approach in the continuous monitoring of natural waters (and/or effluents). It concerns physico-chemical and biological multi-sensors where information will be analysed in a common way by an expert system. Each type of organism, possibly each species, will have a specific sensivity for various pollutants and/or for polluant mixture (see figure 1). For example, herbicides will be better detected by algae (Pandard et al., 1993), anti-fouling agent by bivalves (Jenner et al., 1989). For this reason, a priority research action will be to develop a combination of the various types of organisms (which will result in the extension of the range of detected pollutants) in parallel with the analysis of physico-chemical sensors information. As to the biological approach, various types of organisms can be used as fish, bivalves, daphnia, algae, bacteria, ... and also different responses will be analysed as behaviour, physiology, biochemical changes, ... (Kramer and Botterweg, 1989; Thomas, 1997; Van der Schalie, 1986). This will also involve biosensors that use immobilized enzyme cells, mitochondrial and tissue biocatalysts, antigen-antibody inhibitor reactions, ... (Tran-Minh, 1991).

In fact, the idea is to provide the scientific and technical bases for improved water surveillance capable of rapidly indicating quality deterioration. It is also aimed to intregrate several physico-chemical data available from common sensors (pH, temperature, suspended matter, dissolved oxygen, ...) with information of the biological early warning systems and biosensors using several organisms and biological components in a multi-sensor approach. The nature of the retained sensors has to be related to a vulnerability study.

3. Current state of the art

The usefulness of the two approaches (physico-chemical and biological) to assure reliable monitoring of water quality is now generally recognised (Wallwork and Ellison, 1983; Botterweg et al., 1989). Scientific literature abounds in biological early warning systems and biosensors the diversity of which is attested not only by the variety of biological reagents used but also by the number of principles developed (Thomas, 1996; Thomas and Terver, 1994). Some biomonitors record the respiration activity by measuring the oxygen concentration in very different test organims, such as bacteria (Solyom, 1977), algae (Geller and Mäckle, 1977), invertebrates (Heinis and Swain, 1986). Several monitors evaluate the light emission of bacteria (Bulich, 1979), the photosynthetic activity (Pandard and Vasseur, 1992) or the fluorescence production (Benecke et al., 1982) of algae. Other examples include the use of crustacean locomotion (Knie, 1988), mussel valve activity (Kramer et al., 1989), of fish behavioural (Besh et al., 1977; Poels, 1977), physiological (Morgan and Kühn, 1974) and electric responses (Rauch, 1980; Thomas et al., 1996) in the development of early warning systems. In spite of the proliferation of biomonitoring systems (see Table 1), much remains to be done in terms of research to grapple with a number of existing problems.

Crucial among these is the determination of water quality thresholds and pollution detection time in keeping with drinking water quality standards. In addition, the various biological monitoring systems are at different development levels with not always a validation in situ. For the sake of reliability, it is also crucial to distinguish between biological responses caused by normal variations of environmental parameters and those caused by the presence of toxic substances. For example, the influence of temperature variations on biological information is a common problem facing all biological early warning systems. Like this, Bulich (1979) and Vasseur et al. (1986) use the light ouput of luminescent bacteria for determining the toxicity of water samples. It was shown that light emission is temperature dependent and consequently it was necessary to control very precisely the water temperature. The same problem appears in poïkilothermic electric fish, as Apteronotus albifrons (Gymnotiformes). In these organisms, we can exploit the Electric Organ Discharge Frequency, a physiological response affected by the water temperature (Thomas et al., 1997). In this context, the development of methods integrating several physico-chemical and biological sensors could be a promising means and proposes a completely new approach as an integrated early warning system.

4. Further research needs

The outcome of the further research would be the creation of a highly sensitive monitoring approach (with physico-chemical and biological sensors) for aquatic environment quality surveillance. It will aim specially at improving biological early warning systems which are required by national and international commissions and organizations for the implementation of monitoring programmes. This is all the more important since the number of new chemical products put on the market increases year by year (Chemical Abstracts).

As to the scientific, technical and economic arguments, several points can be listed:

- only a limited number of compounds can be detected with physico-chemical sensors;
- there is a complementarity of physico-chemical and biological approaches;
- detection of possible effects of complex mixture of pollutants to different species, for a global approach;
- detection of sudden changes in river water quality and safeguarding waterwork intakes against toxic events;
- a relatively cheap method of broad-spectrum detection with the biological approach;
- the reliability of a multi-system approach.

The information provided by such an integrated method can also be useful for the maintenance of the general system. Local authorities, water agencies and national environmental authorities have already expressed their interest for such a project. On the other hand, biological monitoring is a relatively cheap method of broad-spectrum detection which can support early warning and survey programmes of organizations such as the International Rhine and Danube Commissions, ...

Figure 1 : Comparison of the sensitivity of biosensors applied to water quality control (modified after Osbild et al., 1995).

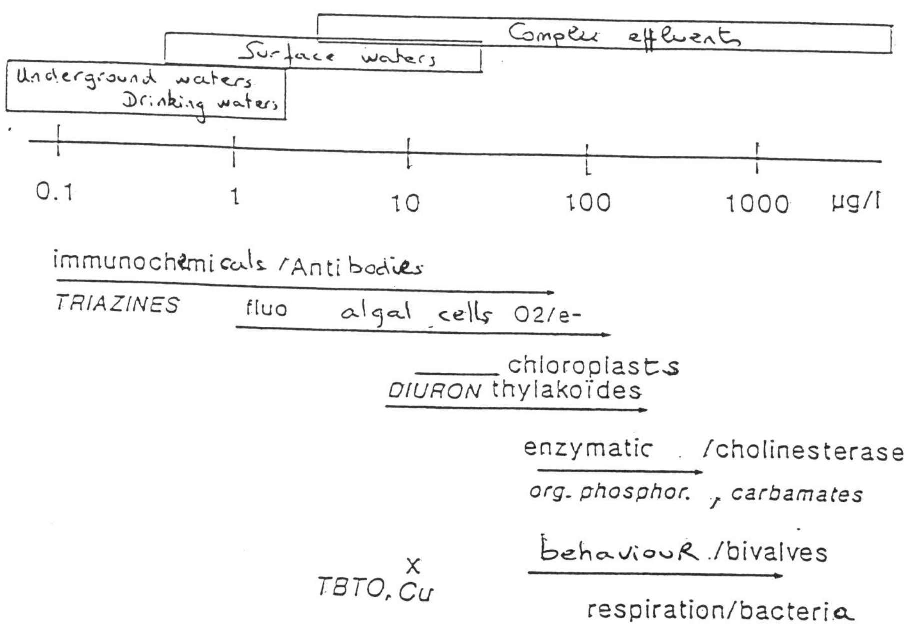

Table 1 : Overview of the automated biological early warning systems or biosensors (modified after WIR Bund / Länder - Projektgruppe "Wirkungstest Rhein", 1994).

Biological reagents	Test systems	Biological components or test organisms	Test parameters
ENZYME	Enzymatic biosensors	Tyrosinase Cholinestérase Cytochrome oxydase Acétylcholinestérase Butyrylcholinestérase	Enzyme activity Enzyme activity Enzyme activity Enzyme activity Enzyme activity
ANTIGEN-ANTIBODY	Immunosensors		
BACTERIA	Toxiguard Biox 1000T Toxalarm Stiptox-norm EuCyano bacteria electrode RODTOX Lumino 2000 RBT Biotoximeter Auto-Microtox	Biofilm Biofilm Pseudomonas putida Pseudomonas putida Escherichia coli Activated sludge Photobacterium phosphoreum Photobacterium phosphoreum Photobacterium phosphoreum Photobacterium phosphoreum	Respiration: O2 consumption Respiration: O2 consumption Respiration: O2 consumption Respiration: O2 consumption Respiration: electron transport Respiration inhibition Inhibition of nitrifying ability Light emission Light emission Light emission Light emission
ALGAE	FluOx test system Biosens algae toximeter IFW fluorometer DF algae test Biotoxitel EuCyano bacteria electrode	Chlamydomonas reinhardii Chlamydomonas reinhardii Chlamydomonas reinhardii Chlamydomonas reinhardii Chlorella vulgaris Synechococcus sp.	Photosynthesis: fluorescence and O2 development Photosynthesis: spontaneous fluorescence Photosynthesis: spontaneous fluorescence Photosynthesis: delayed fluorescence Photosynthesis Photosynthesis: electron transport
HIGHER PLANTS	Protoplast biotest	Vicia faba	Photosynthesis: O2 development

Biological reagents	Test systems	Biological components or test organisms	Test parameters
ANIMALS	Dynamic daphnia test	Daphnia magna	Swimming activity
	Dreissena monitor	Dreissena polymorpha	Valve movement
	Mosselmonitor	Dreissena polymorpha	Valve movement
	Aqua-Tox Control	Leuciscus idus melanotus	Swimming activity
	Fish warning test	Leuciscus idus melanotus	Swimming activity
	Behavioural fish test	Leuciscus idus melanotus	Swimming behaviour
	WRc fish monitor	Oncorhynchus mykiss	Ventilation: gill beat frequency
	Gymnotox	Apteronotus albifrons	Electric organ discharge frequency

REFERENCES

1. Benecke, G., Falke, W. and Schmidt, C., 'Use of algal fluorescence for an automated biological monitoring system', Bull. Environ. Contam. Toxicol. 28, 385-395, 1981.
2. Besch, W.K., Kemball, A., Meyer-Waarden, K. and Scharf, B, 'A biological monitoring system employing rheotaxis of fish', In: Biological monitoring of water and effluent quality, J.Jr. Cairns, K.L. Dickson & G.F. Westlake (Eds.), ASTM STP 607, American Society for Testing and Materials, 56-74, 1977.
3. Botterweg, J., Van de Guchte, C. and Van Breemen, L.W.C.A., 'Bio-alarm systems: a supplement to traditional monitoring of water quality', H_2O, 22, 778-794, 1989.
4. Bulich, A.A., 'Use of luminescent bacteria for determining toxicity in aquatic environments', In: Aquatic toxicology, L.L. Marking & R.A. Kimerle (Eds.), ASTM STP 667, American Society for Testing and Materials, 98-106, 1979.
5. Geller, W. and Mäckle, H., 'Uberwachung von Roh- und Trinkwasser. Teil 2 : Kontinuierlicher Biotest zur Toxizitätsüberwachung von Trinkwasser', DVGW-Schritenreihe Wasser 14, 173-181, 1977.
6. Heinis, F. and Swain, W. R., 'Impedance conversion as a method of research for assessing behavioural responses of aquatic invertebrates', Hydrobiol. Bull. 19 (2), 183-192, 1986.
7. Jenner, H.A., Noppert, F. and Sikking, T, 'A new system for the detection of valve-movement response of bivalves', Kema Scientific and Technical Reports 7 (2), 91-98, 1989.
8. Knie, J, 'Der Dynamische Daphnientest, praktische Erfahrungen bei der Gewässerüberwaschung', Gewässerschutz Wasser Abwasser 102, 341-356, 1988.
9. Kramer, K.J.M. and Botterweg, J, 'Aquatic biological early warning systems: an overview', In: Bioindicators and environmental management (1991), D.W. Jeffrey & B. Madden (Eds.), 6th International Bioindicators Symposium, Trinity College, Dublin, 23-28th September 1990, Academic Press, UK, 95-126, 1990.
10. Kramer, K.J.M., Jenner, H.A. and De Zwart, D, 'The valve movement response of mussels: a tool in biological monitoring', Hydrobiologia 188/189, 433-443, 1989.

11. Morgan, W.S.G. and Kühn, P.C., 'A method to monitor the effects of toxicants upon breathing rate of largemouth bass (Micropterus salmoides, Lacépède)', Wat. Res. 8, 67-77, 1974.
12. Osbild, D., Babut, M. and Vasseur, P., 'Etat de l'art. Les biocapteurs appliqués au contrôle des eaux', Rev. Sci. Eau 8, 505-538, 1995.
13. Pandard, C.P. and Vasseur, P., 'Biocapteurs pour le contrôle de la toxicité des eaux : application des bioélectrodes algales', Rev. Sc. Eau 5, 445-461, 1992.
14. Pandard, P., Vasseur, P. and Rawson, D.M., 'Comparison of two types of sensors using eukaryotic algae to monitor pollution of aquatic systems', Wat. Res. 27 (3), 427-431, 1993.
15. Poels, C.L.M., 'An automatic system for rapid detection of acute high concentrations of toxic substances in surface water using trout', In: Biological monitoring of water and effluent quality, J.Jr. Cairns, K.L. Dickson & G.F. Westlake (Eds.), ASTM STP 607, American Society for Testing and Materials, 85-95, 1977.
16. Rausch, E., 'Einrichtung zur Überwaschung des Wasserqualität mithilfe elektrischer Fische', Patent Application, publication nr. and date: DE 29.06.884, 28.08.1980.
17. Solyom, P., 'Industrial experiences with toxiguard, a toxicity monitoring system', Prog. Wat. Tech. 9 (5/6), 193 –198, 1977.
18. Thomas, M. and Terver, D., 'Le recours biologique pour la surveillance de la qualité des eaux', Ichtyophysiologica Acta 17, 167-188, 1994.
19. Thomas, M., Florion, A. and Chrétien, D., 'A new warning biomonitor using a weakly electric fish, Apteronotus albifrons (Gymnotiformes) and the effect of temperature on the bioelectric responses', Environ. Mon. Assess., in press, 1997.
20. Thomas, M., Florion, A., Chrétien, D. and Terver, D, 'Real-time biomonitoring of water contamination by cyanide based on analysis of the continuous electrical signal emitted by a tropical fish, Apteronotus albifrons', Wat. Res. 30 (12), 3083-3091, 1996.
21. Thomas, M., 'Biodétection en temps réel. Contribution à la réalisation d'un nouveau système de détection biologique de la pollution des eaux de surface, fondé sur l'exploitation des décharges auto-stimulatrices d'un poisson électrique tropical : Apteronotus albifrons (Apteronotidae - Gymnotiformes)', Thèse de doctorat de l'Université Henri Poincaré, Nancy 1, 1-240, 1996.
22. Thomas, M., 'Etat de l'art sur les systèmes biologiques d'alarme précoce utilisés dans la surveillance des eaux brutes destinées à la production d'eau potable', In: 77e Congrès de l'AGHTM, Association Générale des Hygiénistes et Techniciens Municipaux, La Rochelle, 120-142, 12-16 mai 1997.
23. Tran Minh, C., 'Les biocapteurs : principes, construction et applications', Edition Masson, Paris, 1-158, 1991.
24. Van Der Schalie, W.H., 'Can biological monitoring early warning systems be useful in detecting toxic materials in water ?', In: Aquatic toxicology and environmental fate, Ninth volume, ASTM STP 921, T.M. Poston & R. Purdy, (Eds.), American Society for Testing and Materials, Philadelphia, 107-121, 1986.
25. Vasseur, P., Férard, J.F., Rast, C. and Weingertner, P., 'Le test Microtox et le contrôle de la qualité des eaux', J. Fr. Hydrol. 17 (2), 153-162, 1986.
26. Wallwork, J.F. and Ellison, G, 'Automated sensor systems for water resource pollution warning and treatment process control', Aqua 6, 313-320, 1983.

27. WIR Bund / Länder - Projektgruppe "Wirkungstest Rhein", 'Continuous biotests for water monitoring of the river Rhine', Summary, Recommendations, Description of Test Methods. Umweltbundesamt, Berlin, UBA-Texte, 1-54, 1994.

IMPACT ON THE QUALITY OF THE INDUSTRIAL SIDERURGICAL WATERS BY REPLACING THE CLASSICAL BLAST FURNACES' BF/LDAC BASIC OXYGEN FURNACES' BOF FILE (WET GAS CLEANING) THROUGH THE ELECTRIC ARC FURNACES' EAF FILE (DRY GAS CLEANING)

G. DIDERICH, Profil ARBED SA, Contrôle et Gestion Environnement, Esch-sur-Alzette, Luxembourg

1. Introduction

The plants of ProfilARBED and ARES in Luxembourg represent two business units of the ARBED group that, with production capacities of 11 millions tons of steel per year, is the 4th iron and steel industry in Europe. They employ in Luxembourg about 5500 persons and have a production capacity evaluated at 3 millions tons of steel per year. ProfilARBED is focused on long heavy products. Long light products are concentrated at ARES, Esch-Schifflange and Rodange.

During the period 92/94 ARBED has decided to pass, in the Luxembourg plants, from the classical file (sinter plant, blast furnaces BF, LDAC basic oxygen furnaces BOF) to the all electrical file (electric arc furnaces EAF, ladle heating furnaces LHF, continuous casting machines CCM, walking beam furnaces WBF). Thus, the classical file will be replaced by three electric arc furnaces in the plants of Esch-Belval, Differdange and Esch-Schifflange. This structural change will be finished in the second half of 1997.

2. Target of the report

The present report has as target to show the impact on the quality of industrial siderurgical water by changing the classical file, with wet gas cleaning, through the all electrical file, with dry gas cleaning, at the plant of ProfilARBED, Esch-Belval.

3. Water circuits for the classical file with dry gas cleaning

3. 1. Situation concerning water supplies

All the Luxembourg plants of ARBED are situated far away from the great rivers of the region and are thus largely limited concerning the water supplies and the industrial water discharge. Therefore, they are oriented from their beginning, as far as possible, on closed circuits. Low water consumptions' are the consequence.

3. 2. Potential pollutants of the siderurgical water

A great variety of pollutants, in variable concentrations, may be found in the siderurgical waters of the different kinds of production installations. These waters are either washing waters of off-gases, process waters or cooling waters.

The waters of the overflow of the internal circuits flow in the rolling mills' pond or in the reserve pond. They may hold suspended solids, hydrocarbons (oil/grease), heavy metals (Zn, Pb, Cd, Fe, Cu ....) or other pollutants (BOD-5, COD, NO_2, NH_4.........).

The siderurgical waters are checked by a systematic follow-up of the water treatments of the different internal circuits and by a day per day monitoring of the water quality of the effluent.

3. 3. Monitoring of the waters of the internal circuits and of the effluent

In order to follow continuously the quality of the siderurgical waters a systematic check is realized on the waters of the internal circuits to adapt respectively to improve the physical/chemical treatment modes of these waters and to supervise the effluent of the plant.

To check the effluent's water a continuous sampling equipment, corresponding to the legislation, is installed. Furthermore, on-line measurements of pH, temperature and conductivity of the effluent are also installed.

The sampling frequencies and the chemical analysis are realized according to the working permit of the plant. In order to calculate the pollution charge leaving the plant a continuous measurement of the water flow is also put in the effluent.

The waters of the internal circuits are sampled manually and analysed according to a procedure defined individually for every circuit.

The management of the results with regard to the waters of the effluent and of the internal circuits is carried out by the sector " Contrôle et Gestion Environnement " of ProfilARBED. The results determined on the water of the effluent are forwarded every month to the national Administration of Environmental Affairs.

3. 4. Water supplies and water discharges

The only water supplies of the plant of ProfilARBED, Esch-Belval are a small brook " Wenschel ", the flow of which varies between 20 and 200 m^3/h, and the water effluent of an external communal mechanical water treatment installation of the commune of " Belvaux " STEP, the flow of which varies between 50 and 200 m^3/h. Besides these supplies, the plant collects a great deal of rain water, surface water and sanitary water and is also obliged to take intercommunal water.

The water's discharge is composed exclusively by the overflow of the reserve pond that flows directly into the brook " Dipbach ", the flow of which varies in large proportions, 0-50 m³/h during dry periods, and up to 500 m³/h during important rain falls. Normally, the overflow of the rolling mills' pond passes only into the reserve pond by three sewer pipes. Floating dams hold back hydrocarbons of the rolling mills that have passed accidentally in the rolling mills' pond. If necessary, they are recuperated by pumping and eliminated by an aggregated firm.

Both ponds, that make up the water reserve of the plant, have as other function to allow the suspended solids to deposit. Cleaning of these ponds by an aggregated firm is done periodically

The cooling water supplied to the blast furnaces comes from the two blast furnaces' pools and goes back to them after utilization.

The water for the rolling mills is supplied by the rolling mills' pond and goes back to this same pond after clarification and disoiling in a modern thickener where mill scale sludge and used oils and greases are recovered.

The water balance for 1996 is given in the table below.

Specific water consumption
(Input water-Output water)/steel production: (4.775.000-1.850.000)/1.168.068 = 2,5 m³/t steel

Input	**m³/year**
" Wenschel "	2.700.000
STEP	795.000
Intercommunal water	900.000
Rain water/sanitary water	380.000
Total	4.775.000

Outtput	**m³/year**
Effluent plant, " Dipbach "	1.850.000

This specific water consumption can be considered as one of the lowest in the world's iron and steel industry. It could be realized by the systematic development of closed or semi-closed circuits and air-tight circuits with high recycling rates.

3. 5. *Internal circuits*

3. 5. 1. Closed and semi-closed circuits

3. 5. 1. 1. Washing circuit of blast furnaces' gas

After primary purification in the pots and cyclones, the blast furnaces' gas contains about 450 mg/m^3 dust. The final purification is done in Venturi washers of a wet Bischoff dedusting system.
The water recuperated at this washing is treated in a thickener. Acid out of the blast furnace (pH 4-6), the water is neutralized to pH 8,2-8,7 by regulation through addition of soda. This allows a quasi quantitative precipitation of the metals Zn, Pb, Cd soluble at pH < 8. A floculation agent is added in the thickener to increase the decanting speed of the sludge. The sludge, with high zinc content, and not recyclable in the blast furnace, is pumped to a decantation basin until the disponibility of an economic justifiable valorization technology.
The water supply comes from the reserve pond and the purge is evacuated through the blast furnaces' pools into the same pond.

3. 5. 1. 2. Granulation circuit of blast furnaces' slag

The installations have a granulation capacity of about 1.000.000 t slag/an. The water supply necessary to compensate the loss of water contained in the slag and the evaporated water, evacuated through a chimney, comes from the blast furnaces' pools. The purge is evacuated into the reserve pond.

3. 5. 1. 3. Washing circuit of the LDAC basic oxygen furnace's gas

After combustion of the converter gases, charged with CO, in a smoke gas boiler, that recovers energy, the off-gas is washed by Venturi washers of a Bischoff dedusting system.
The sludge issued from this operation is collected in a pre-thickener (coarse) and in a thickener (fine). The coarse sludge is recycled through the sinter plant and the fine sludge, not recyclable at the blast furnace, is dumped until the disponibility of an economic justifiable valorization technology.
To avoid lime deposits in the circuit, additions of dispersants and Na_2CO_3 are carried out to soften the water of the circuit. The precipitation of $CaCO_3$ is generated in the thickener and the $CaCO_3$ is evacuated with sludge.
The water supply of this circuit comes from the rolling mills' pond. The purge is evacuated through the circuit of the rolling mills into this same pond.

3. 5. 1. 4. Cooling circuit of the rolling mills

Out of the rolling mills, the raw waters contain 150-450 mg/l suspended solids (mill scale and mill sludge) and 10-30 mg/l hydrocarbons. They are introduced by gravity in a mechanical pre-thickener, equipped with 8 arms capable of bringing the mill scale < 0,2 mm (coarse) to a classification installation related with the pre-thickening compartment.
The classification installation gets the mill sludge < 0,2 mm (fines), recuperated at the underflow of the thickener-disoiler. These fines, charged with oil and grease, are pumped to a decantation basin where they are extracted periodically in order to be stocked until the disponibility of an economic valorization technology. The mill scale > 0,2 mm is evacuated by a skip. Due to the low oil/grease content it can be recycled in the sinter plant.
The pre-decanted water is evacuated to the thickener-disoiler equipped with a double blade-scraper allowing the sweeping of the entire surface of the thickener and the evacuation of the oil/grease residues in a post-separation tank. These residues are eliminated by an aggregated firm. The cleaned water returns to the classification installation.
The effluent of the thickener-disoiler has among 20 and 50 mg/l suspended solids and less than 5 mg/l hydrocarbons. The cleaned water flows afterwards into the rolling mills' pond and then in the reserve pond. The overflow of this pond represents the effluent of the plant, the " Dipbach ".
The water effluent standards of the National Environmental Administration are shown in Table 1. They are respected with some rare minor exceptions (Zn, NO_2, BOD-5).
For all these circuits the recycling rate is higher than 90%.

3. 5. 2. Hermetically closed water circuits

Besides the closed and half-closed water circuits, different installations of the BF and the BOF are equipped with hermetically closed water circuits. They are run with demineralised water that is treated with additional products to protect the walls of the water mains from corrosion and deposits.

3. 6. Impact of the different water circuits on the chemical quality of the effluent of the plant

3. 6. 1. Contamination by heavy metals

An uncontrolled purge respectively overflow of the circuits and faulty BF and BOF water treatment installations (thickeners) may breed higher values of Zn and/or Pb of the water of the reserve pond and thus of the effluent of the plant. An overtaking of the standards may be the consequence

3. 6. 2. Contamination by hydrocarbons

Only a faulty rolling mills' water treatment installation (thickener) may breed higher values of hydrocarbons of the water of the rolling mills' pond. In such a situation, an oil pumping operation is executed. Besides floating dams are installed to avoid the discharge of hydrocarbons over the reserve pond into the effluent. Pollution risks are thus limited.

3. 6. 3. Contamination by BOD-05, NH_4, NO_2

The water supplied by the mechanical communal water treatment installation of the commune of " Belvaux " may cause sporadic increases of BOD-5, NH_4 and NO_2 of the effluent of the plant.
This situation is however of minor importance. Indeed, an increase of these parameters is on one hand of no influence on the internal water circuits and and on the orher hand it should be emphasized that the effluent " Dipbach " is treated some km further away in a biological communal water treatment installation.

3. 7. Chemical quality of the effluent of the plant

During 1996, due to a systematic monitoring of the water quality of all the circuits, the mean values of the parameters of the effluent did never show significant overtaking of the standards.

However, it should be pointed out that the waters of the two ponds are characterized by a higher salinity (conductivity) and a hardness higher compared to the hardness of intercommunal water. These values are originated by the salty overflow of the wet cleaning installations of the BF and the BOF. Due to dilution by rain water the values are lower in winter than in summer.

Considering the mean values, the standards for heavy metal and hydrocarbons are largely respected. This result is explained by the continuous and systematic monitoring.of the water quality of all the circuits.

Only punctual small increases of the parameters Zn, BOD-5, NH_4 and NO_2 are observed especially in the summer time. They are caused by an uncrontolled overflow of the blat furnaces' circuit and by the continuous and necessary supply of the water from the mechanical urban water treatment installation of the border commune " Belvaux " of the plant.

4. Water circuits of the all electrical file with dry gas cleaning (Fig. 2)

4. 1. Closed circuits

The waters of the new siderurgical installations (EAF, LHF, CCM, WBF) are run in closed and semi-closed circuits. The water supplies come from the reserve pond, from the rolling mills' pond and from the intercommunal water network.

These circuits are treated with chemical agents to avoid corrosion and encrustation.

Furthermore, for the CCM, due to an acidification caused by the use of fluorspar as lubrification agent, the water of this circuit has to be neutralized by NaOH.

The rolling mills' circuit is not at all changed. The water continues to be treated in a clarification-disoiling installation that is of best available technology.

4. 2. Hermetically closed cooling circuits

Besides the closed and the semi-closed circuits, the new installations are equipped with hermetically closed circuits. The water of them is systematically treated by chemical agents to avoid corrosion and calcareous deposits. The water losses are compensated by water supplied from the intercommunal network.

4. 3. Chemical quality of the effluent of the plant

The waters of the overflow of the different circuits of the new file will no more be charged with heavy metals and the salinity of them will be largely reduced. Therfore, it is evident that the chemical quality of the reserve and the rolling mills' ponds will be improved by the replacement of the BF/BOF file through the EAF file.

The only possible remaining pollution will be an uncontrolled leakage of hydrocarbons. However, as the rolling mills' water treatment can be considered as best available technology, the risks of higher hydrocarbons in the effluent are limited. The supplementary dams installed in the rolling mills' pond are to be considered as a further security.

4. 4. Water needs for the EAF file

The theoretical needs of water to compensate the evaporation and the necessary purge are estimated at 1.700.000 m^3/year. This corresponds, for an aimed production of 1.000.000 t steel/year, to a specific water consumption of 1,7 m^3/ t steel, value diminished by about 30% compared to the value measured for the BF/BOF file.

5. Conclusions

The replacement of the BF/BOF file (wet gas cleaning) at the ProfilARBED, Esch-Belval steel plant by the EAF file (dry gas cleaning) will have the following positive consequences :

- a decrease from up to 30% of the specific water consumption due to the installation of closed, semi-closed and hermetically closed water circuits,
- an roughly zero risk of heavy metal pollution in the effluent of the plant,
- a progressive decrease of the salinity (conductivity) and the hardness of the water of the reserve and the rolling mills' ponds and thus a drastic improvement of the chemical quality of the effluent's water,
- a limited risk of hydrocarbon pollution due to a rolling mills' water treatment installation thas is of best available technology,
- the systematic and continuous monitoring of the quality of the waters of the internal circuits, the ponds and the effluent is a further assurance to avoid pollution in the effluent, location of assessment of the water quality by the National Environmental Administration.

Table 1 : Water effluent standards for ProfilARBED, Esch-Belval and Differdange and for ARES, Esch, Schifflance

Parameter	Unit	Standard
pH		6.5-9
Temperature	° C	< 30
Suspended solids	mg/l	< 30
Conductivity	µS/cm	p.i.
Fe	mg/l	< 2
Zn	mg/l	< 2
Pb	mg/l	< 0,5
Cd	mg/l	< 0,1
Cu	mg/l	< 0,5
As	mg/l	< 0,1
Hg	mg/l	< 0,01
Cr tot.	mg/l	< 0,5
Cr VI	mg/l	< 0,1
NO3 (drinking water)	mg/l	< 50
NO2	mg/l	< 1
CN free	mg/l	< 0,1
Phenol	mg/l	< 0,1
Cl organic	mg/l	< 0,1
BOD-5	mg/l	< 20
COD	mg/l	< 100
NH4	mg/l	< 10
Hydrocarbons	mg/l	< 5

Water sampling by continous automatic equipments
Continuou registration of water flow, pH, temperature and conductivity

Figure 1 : Industrial water circuits of the ProfilARBED plants, Esch-Belval (PA-EB)
Classical file, blast furnaces BF/LDAC steel plant BOF (wet cleaning of off-gasses)

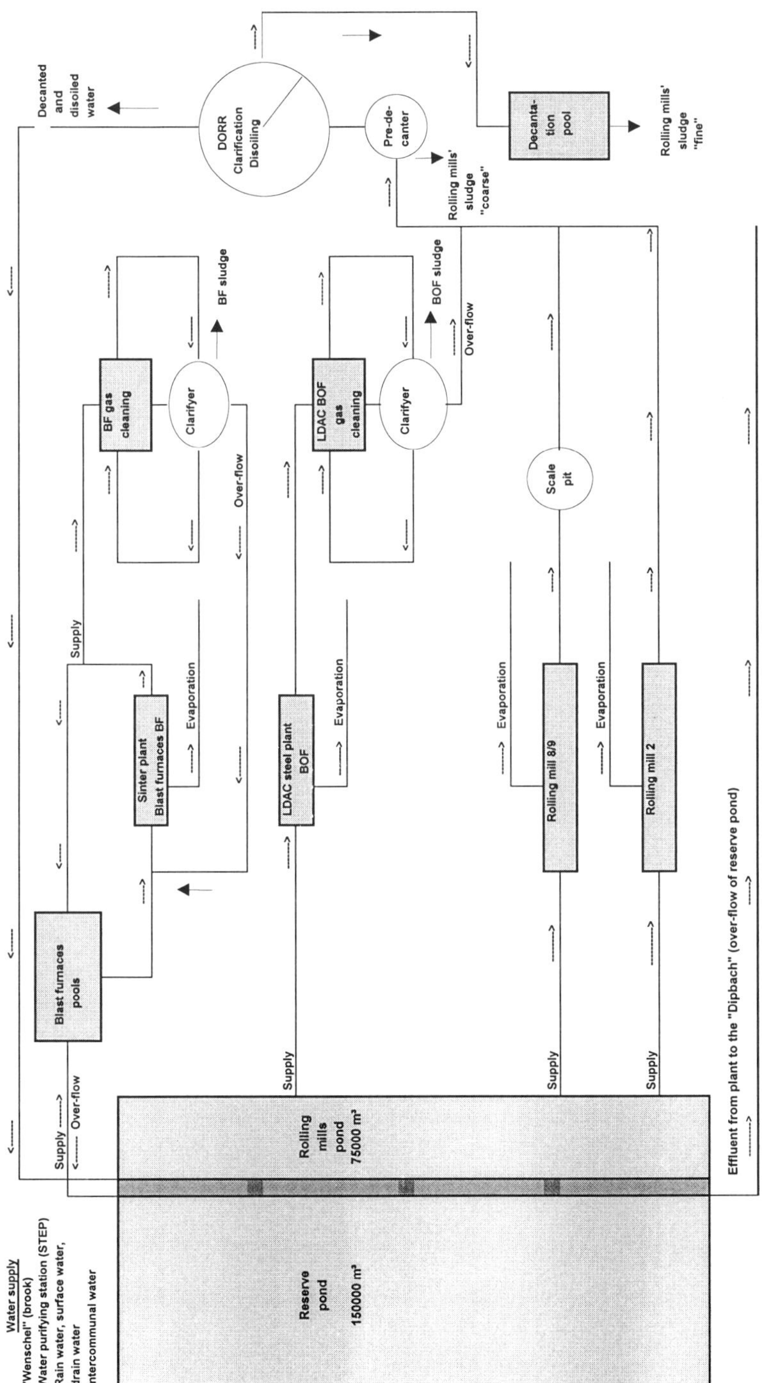

Figure 2 : Industrial water circuits of the ProfilARBED plants, Esch-Belval (PA-EB) All electrical file, electrical arc furnance EAF (dry cleaning of off-gases)

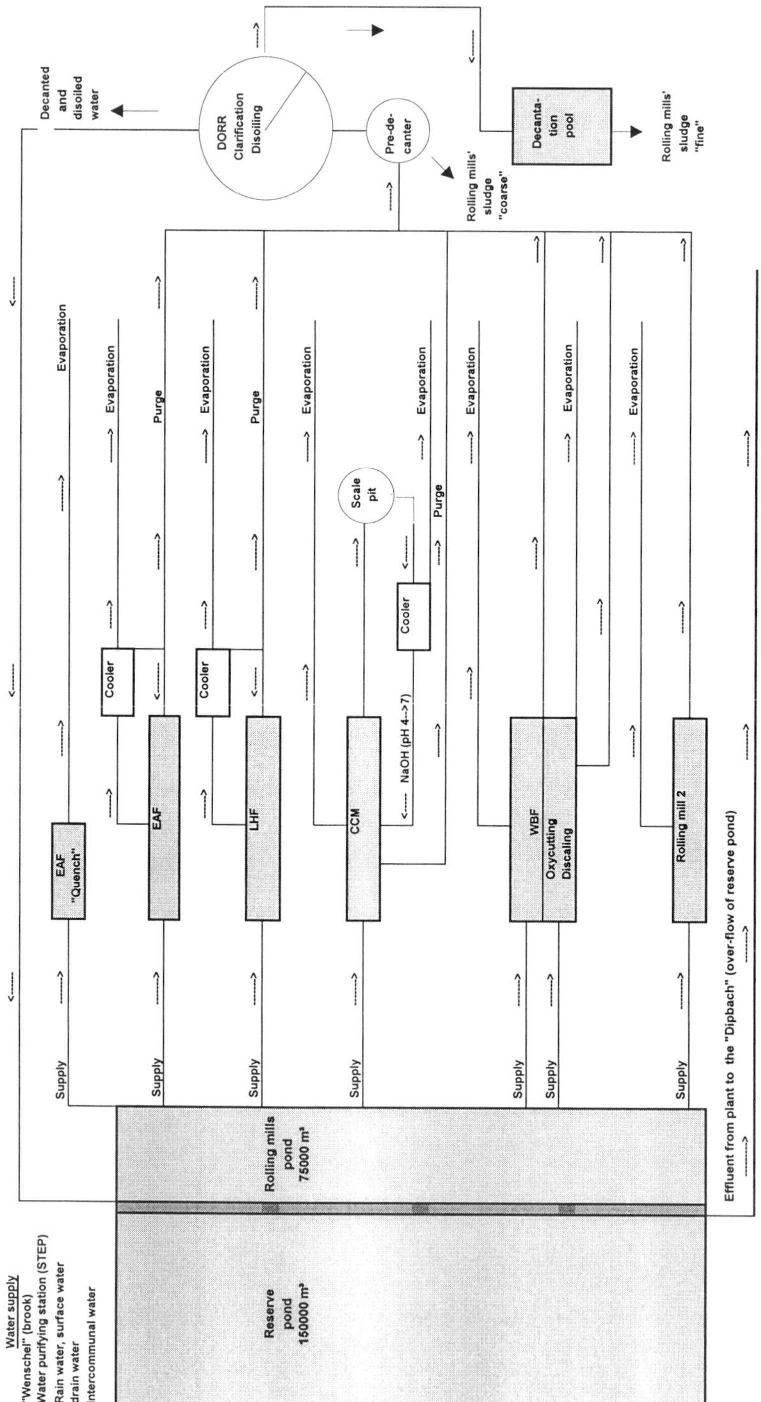

V. WORKSHOP PROGRAMME

European Workshop on Standards, Measurements and Testing for the Monitoring of Water Quality : the Contribution of Advanced Technologies, Nancy (France), 29-31 May 1997

Thursday 29 May 1997

Welcome by :
F. COLIN, Université Henri Poincaré, Nancy I,
Chairman, Scientific Commitee of NAN.C.I.E., Nancy, (F)
and opening address by :
P. QUEVAUVILLER, European Commission, DGXII,
Brussels (BE)

Introduction to the Workshop by F. COLIN

- Aims of the Workshop
- Working method
  - identification of needs
  - identification of technical opportunities
  - matching needs and opportunities

SESSION 1

End User's view : state of the art, identification and prioritisation of needs.

State of the art of Water Quality Monitoring :

- application fields
- measurement aims
- measured parameters and considered concepts
- present performance
- main drawbacks and limitations

Identification and prioritisation of technical and regulatory needs

- classical parameters : required performance
- new concepts to consider, new parameters to be measured

Round-table A : Characterisation of clean waters
Chair : J. BENOLIEL (P)
Rapporteur : T. H. M. NOIJ (NL)

- Surface and underground natural waters
- Drinking water
- Industrial cooling and process waters

Round-table B : Characterisation of waste waters
Chair : A. KATSIRI-KOUZELI (GR)
Rapporteur : H. WACHEUX (F)

- Domestic waste-waters
- Industrial waste-waters
- Waste-water sludge

SESSION 2

Presentation of selected advanced technologies
Chair : J. GUTIERREZ (SP)
Rapporteur : D. KLOCKOW (D)

- introductory presentation : F. COLIN (F)
- optical technologies : P. SCULLY (UK)
- electrochemical sensors : P.PAYNE (UK)
- microbiological sensors : P. VASSEUR (F)
- biochemical sensors : D. CULLEN (UK)

Friday 30 May 1997

SESSION 3

Advanced Technologies for Classical parameters : improvement of existing Methods or Development of Alternative Techniques ?
Chair : G. FREGO (I)
Rapporteur : A. KATSIRI KOUZELI (GR)

General discussion on :

- what possibilities ?
- what adaptation problems ?
- specific advantages : autonomy, miniaturisation, performance, cost, ..

Round-table C : Global parameters
Chair : J-L CECILE (F)
Rapporteur : D. CULLEN (UK)

- pH, redox, conductivity
- suspended matter
- total organic load
- biodegradable organic matter
- toxicity
- odours

Round-table D : individual chemical species
Chair : T. H. M. NOIJ (NL)
Rapporteur : P. PAYNE (UK)

- dissolved gas (O_2, CO_2)
- dissolved solids (anions and cations)
- N species
- P species
- organic micropollutants (pesticides, BTEX, HPA, PCB, ...)
- inorganic, micropollutants (metals and metalloids)

SESSION 4

Advanced Technologies for New Parameters and Measurement Concepts
Lecturer : M. JAUZEIN (F)
Rapporteur : M. SCOTT (UK)

- chemometrics multivariate techniques : B.G.M. VANDEGINSTE (NL)
- sensor arrays
- virtual parameters
- highly specific measurement systems
- integrated data treatment

Round-table E : multivariate techniques for behavioural characterisation of waters
Chair : B.G.M. VANDEGINSTE (NL)
Rapporteur : J. RAGOT (FR)

Round-table F : highly specific techniques for individual chemical species
Chair : D. KLOCKOW (D)
Rapporteur : G. ORELLANA (SP)

SESSION 5

Round-table on the theme "Needs and Possibilities for the Application of Advanced Technologies, Establishing Priorities :
Presentation by rapporteurs of sessions 1 to 4 and
chairmen of Round Tables

Conclusions regarding the SM&T Programme
P. QUEVAUVILLER, European Commission, DG XII, Brussels

VI. LIST OF PARTICIPANTS

Mrs. BENOLIEL Maria Joao
EPAL - Empresa Potuguesa das Aguas Livres SA
Rua do Alviela 12
1170 Lisboa Codex
PORTUGAL
Telephone : (351) 1 815 33 31
Fax : (351) 1 813 99 74

Mr. CECILE Jean-Luc
GEMCEA
149, rue Gabriel Péri
54 500 Vandoeuvre
FRANCE
Telephone : (33 3) 83 57 88 11
Fax : (33 3) 83 53 45 02

Prof. COLIN François
Workshop Chairman
I.R.H.
11 bis, rue Gabriel Péri
54 500 Vandoeuvre
FRANCE
Telephone : (33 3) 83 50 36 00
Fax : (33 3) 83 50 36 99

Dr. CULLEN David C.
Lecturer in Biophysics & Biosensors
Cranfield Biotechnology Centre
Cranfield University
Cranfield
UNITED KINGDOM
Telephone : (44) 1234 750907
Fax : (44) 1234 754339
Email : d.cullen@cranfield.ac.uk

Mr. DIDERICH Georges
Profil ARBED - Site d'Esh-Belval
L-4008 Esch sur Alzette
LUXEMBOURG
Telephone : (352) 5550 22 60
Fax : (352) 5550 34 83

Mr. FREGO Giuseppe
ENEL - Residue Valorisation and Treatment Research Centre
via dalmazia 21/C
74100 BRINDISI
ITALY
Telephone : (39) 831 517 351
Fax : (39) 831 517 305

Mr. GUTIERREZ Monreal Javier
Consejo Superior de Investigaciones Científicas
Centro de Technologias Físicas "Leonardo Torres Quevedo"
Instituto de Física Aplicada
Calle Serrano, 144
ES-28006 Madrid
SPAIN
Telephone : (341) 561 88 06
Fax : (341) 411 76 51
Email : ltqgm10@fresno.csie.es

Mr. JACOBSEN Hans Svankjaer
VKI - Institute for Water Environment
Gustav Wieds Vej 10
DK-8000 Arhus C
DENMARK
Telephone : (45) 86 20 20 00
Fax : (45) 86 19 75 11

Mr. JAUZEIN Michel
IRH-Environnement
11, bis rue Gabriel Péri
54 500 Vandoeuvre
FRANCE
Telephone : (33 3) 83 50 36 00
Fax : (33 3) 83 50 36 99

Dr. KATSIRI-KOUZELI Alexandra
National Technical University of Athens
Iroon Polytechniou 5
GR-15 773 Zographou - Athens
GREECE
Telephone : (301) 7722 797
Fax : (301) 7722 899

Prof. Dr. KLOCKOW Dieter
Institut für Spektrochemie und Angewandte Spektroskopie - ISAS
Bunsen-Kirchhoff-str 11,
D-44013 Dortmund
GERMANY
Telephone : (49) 231 1392 103
Fax : (49) 231 1392 300

Mr. MARTENS Liévin
APPLITEK NV/SA
K Picquelaan 84
B 9800 Deinze
BELGIUM
Telephone : (32) 9 386 34 02
Fax : (32) 9 386 72 97

Dr. Ir. NOIJ Theo H.M.
Head Analytical Method
Development Section
Chemistry and Biology Department
KIWA N.V. - Research and Consultancy
Groningenhaven 7 - P.O. Box 1072
3430 BB Nieuwegein
THE NETHERLANDS
Telephone : (31) 30 60 69 642
Fax : (31) 30 60 61 165
Email : noij@kiwaoa.nl

Dr. ORELLANA Guillermo
Grupo INTERLAB
c/maria Tubau 4 - 3° A1
E-28050 Madrid
SPAIN

Other address :
Dr. ORELLANA Guillermo
Universidad de Madrid
Faculty of Chemestry
Deptm. of Organic Chemistry
E-28050 Madrid
SPAIN
Telephone : (341) 394 42 20
Fax : (341) 394 41 03
Email : orellana@eucmax.sim.ucm.es

Prof. PAYNE Peter
Department of Instrumentation and Analytical Science, UMIST
PO Box 88
UK-Manchester M60 IQD
UNITED KINGDOM
Telephone : (44) 161 200 4883
Fax : (44) 161 200 4879

Mr. QUEVAUVILLER Philippe
European Commission - SM&T Programme
Rue de la Loi 200
BE-1049 Bruxelles
BELGIUM
Telephone : (32) 2 296 33 51
Fax : (32) 2 295 80 72

Mr. RAGOT José
CRAN ENSG
2, avenue de la Forêt de Haye
FR-54516 Vandoeuvre-les-Nancy
FRANCE
Telephone : (33 3) 83 59 59 59
Fax : (33 3) 83 59 56 44

Mr. RAVOIRE François-Daniel
THOMSON-CSF
13, rue Salvadore Allende
FR- 94117 Arcueil
FRANCE
Telephone : (33 1) 41 30 31 13
Fax : (33 1) 14 30 49 18

Mr. SCHMITT Jean-Pierre
International Water Centre - NANCIE
149, rue Gabriel Péri - BP 290
54 515 Vandoeuvre Cedex
FRANCE
Telephone : (33 3) 83 15 87 87
Fax : (33 3) 83 15 87 99
Email : schmittjp@nancie.asso.fr

Mr. SCOTT Michael
SWIG - Process Measurement Technology
27 West Green - Barrington
UK-Cambridge CB2 5RZ
UNITED KINGDOM
Telephone : (44) 1223 870967
Fax : (44) 1223 870967
Email : 106740.1372@compuserve.com

Dr. SCULLY Patricia
Liverpool John Moores University
Byrom St
Liverpool L3 3AF
UNITED KINGDOM
Telephone : (44) 151 231 2361
Fax : (44) 151 298 2624

Dr. TRETTNAK Wolfgang
Joanneum Research
Steyrergasse 17
AT-8010 Graz
AUSTRIA
Telephone : (43) 316 876 220
Fax : (43) 316 876 181
Email : wolfgang.trettnak@joanneum.ac.at

Prof. VANDEGINSTE
UNILEVER Research Laboratorium
Vlaardingen
Olivier van Noortlaan 120 - PO BOX 114
3130 AC VLAARDINGEN
THE NETHERLAND
Telephone : (31) 10 460 53 26
Fax : (31) 10 460 56 71

Prof. VASSEUR Paule
Centre des Sciences de l'Environnement
1, rue des Récollets
57 040 Metz Cedex 1
FRANCE

Telephone : (33 3) 87 75 81 81
Fax : (33 3) 87 78 81 89

Mr. WACHEUX Hervé
Anjou Recherche
109, rue des Côtes
F 78600 Maisons-Laffitte
FRANCE
Telephone : (33 1) 34 93 34 67
Fax : (33 1) 34 93 34 69

Mr. ZAPASNIK Jacek
European and Multilateral Affairs
International Water Centre - NANCIE
149, rue Gabriel Péri - BP 290
54 515 Vandoeuvre Cedex
FRANCE
Telephone : (33 3) 83 15 87 87
Fax : (33 3) 83 15 87 99
Email : zapasnikj@nancie.asso.fr